B. Catoire (Editor)

Electron Spin Resonance (ESR) Applications in Organic and Bioorganic Materials

Proceedings of the First European Meeting January 1990, Lyon, France

Publication partially sponsored by
MRT (French Ministry of Research and Technology)

With 115 Figures and 31 Tables

Springer-Verlag
Berlin Heidelberg NewYork
London Paris Tokyo
Hong Kong Barcelona Budapest

B. Catoire

Groupe d'Application de la Resonance
Paramagnétique Electronique
I.T.F. Lyon
Avenue Guy de Collongue, B.P. 60,
F-69132 Ecully Cedex

Library of Congress Cataloging-in-Publication Data
Electron spin resonance (ESR) applications in oragnic and bioorganic materials
proceedings of the first European meeting, January 1990, (France, Lyon) / B. Catoire (ed.).
"Publication partially sponsored by MRT (French Ministry of Research and Technology)."
ISBN-13: 978-3-642-77216-0 e-ISBN-13: 978-3-642-77214-6
DOI: 10.1007/ 978-3-642-77214-6
1. Electron paramagnetic resonance spectroscopy--Congresses.
I. Catoire, B. II. France. Ministère de la recherche et de la technologie
QP519.9.E433E44 1992
574.19'285--dc20 91-42660

Softcover reprint of the hardcover 1st edition 1992

Typesetting: Camera ready by author

51/3020-5 4 3 2 1 0 – Printed on acid-free paper

Preface

E.S.R. techniques which are mature from a fundamental point of view, now constitute a routine investigation tool in chemistry as well as in biophysics in order to study and to follow the behaviour of radical species.

Among the practical applications, let us mention for instance :

- diffusion phenomena (solid - solid, liquid - solid such as the diffusion of molecules through membranes, films fibers ...),
- study of radical species to improve the elaboration of materials with high added value (composites),
- detection of ionised food,
- liquid crystals, polymers,
- radiochemistry,
- targetting of drugs, toxicology.

A better understanding of basic phenomena allows to optimize industrial products and processes applied as well as in advanced fields as in well established ones.

The high sensivity of ESR Spectroscopy and its derived specific techniques (spin labelling, spin probe, spin trapping ...) offer information on the fine morphological structure of the matter as well as on its behaviour under various treatments otherwise not available.

A Symposium was organized in Lyon (France) in January 1990 to promote the use of ESR. Its originality was to deal with the practical applications of ESR to organic and bioorganic materials. The scope of this meeting was to enlarge the field of application from basic research to more applied subjects and this may concern industrial as well as academic researchers. Moreover, the purpose of this symposium was to promote exchanges between European specialists working in public or private areas.

The French ESR Application Group (GARPE) organizing committee hopes that the communications collected in this volume will be useful.

The GARPE would like to thank all the contributors.

We sincerely acknowledge the MRT (French Ministry of Research and Technology) for partially sponsoring this publication.

Le GARPE
Secretary B. CATOIRE
Institut Textile de France
B.P. 60
F-69132 ECULLY CEDEX

List of Contributors

Dr **J.P. AGNEL**, Département de Physiologie du Végétal et Ecosystèmes, CEN Cadarache, F-13108 Saint-Paul-lez-Durance

O. AKERMANN, INRA, Laboratoire de Nutrition et Sécurité Alimentaire, F-78352 Jouy-en-Josas Cedex

S. AYNIE, Laboratoire de Biologie Physico-chimique, Ecole Nationale Supérieure de Biologie Appliquée à la Nutrition et à l'Alimentation, Campus Universitaire Montmuzard, F-21000 Dijon

Dr **J. BARTHE**, Commissariat à l'Energie Atomique, IPSN/DPT/SIDR, B.P. 6, F-92265 Fontenay aux Roses Cedex

Dr **C.M. BATTESTI**, Laboratoire de Radiochimie des Constituants des Aliments, Département de Physiologie Végétale et Ecosystèmes, CEN de Cadarache, F-13108 Saint-Paul-lez-Durance

Dr **F. BERMANN**, Commissariat à l'Energie Atomique, IPSN/DPT/SIDR, B.P. 6, F-92265 Fontenay aux Roses Cedex

F. BONOSI, Department of Chemistry, University of florence, Via G. Capponi 9, I-50121 Firenze

Prof. **F. BONON**, Universita di Padova, Dipartimento di Chimica Fisica, Padova (Italy)

Prof. **M. BRUSTOLON**, Universita di Padova, Dipartimento di Chimica Fisica, Padova (Italy)

Prof. **M. CAMPS**, Université de Saint-Etienne, 23 Rue du Dr Michelon;F-42100 Saint-Etienne

G. CARUBIA, Department of Chemistry, University of Florence, Via G. Capponi 9, I-50121 Firenze

Dr **B. CASSIER**, Université de Bourgogne, Faculté de Pharmacie, 7 Boulevard Jeanne D'Arc, F-21033 Dijon Cedex

Dr **B. CATOIRE**, Institut Textile de France - Section Lyon, B.P. 60, F-69132 Ecully Cedex

Dr **C. CHACHATY**, Département de Recherches sur l'Etat Condensé, DRECAM/SCM/BP 121, Centre D'Etudes Nucléaires de Saclay, F-91191 Gif sur Yvette Cedex

Prof. **B. CHAILOT**, Université de Bourgogne, Faculté de Pharmacie, 7 Boulevard Jeanne D'Arc, F-21033 Dijon Cedex

Dr **O. CHAIX**, URA 69, CNRS, Institut Dolomieu, Université Joseph Fourier, 15 Rue Maurice Gignoux, F-38031 Grenoble

Dr **H. CHARCOSSET**, Institut de Recherche sur la Catalyse, 2 Avenue A. Einstein, F-69626 Villeurbanne Cedex

Prof. **M. CHE**, Laboratoire de Réactivité de Surface et Structure, URA 1106 CNRS, Université P. et M. Curie, 4 Place Jussieu, F-75252 Paris Cedex 05

Dr C. CHOPARD, Laboratoire de Chimie et Biochimie Pharmacologiques et Toxicologiques, Unité associée au CNRS (URA 400), Université René Descartes, 45 Rue des Saints-Pères, F-75270 Paris Cedex 06

Dr C.L. DAVIES, School of Chemistry and Applied Chemistry, University of Wales College of Cardiff, P.O. Box 912, Cardiff CF1 3TB, UK

Dr P. DERACHE, Université Paul Sabatier, Faculté des Sciences Pharmaceutiques, Laboratoire de Biochimie, 35 Chemin des Maraîchers, F-31052 Toulouse cedex

Dr J.M. DI MEGLIO, UA 792 du CNRS, Laboratoire de Physique de la Matière Condensée, 11 Place Marcelin Berthelot, F-75231 Paris Cedex 05

Dr J.C. EVANS, School of Chemistry and Applied Chemistry, University of Wales College of Cardiff, P.O. Box 912, Cardiff CF1 3TB, UK

Prof. J. FAURE, Physicochimie des Rayonnements, Bat. 350, Université Paris Sud UA 75 du CNRS, F-91405 Orsay Cedex

Prof. A. FEIGENBAUM, INRA, Laboratoire de Nutrition et Sécurité Alimentaire, F-78352 Jouy-en-Josas Cedex

M.J. FILDES, Shell Research Ltd, Thornton Research Centre, P.O. Box 1, Chester CH1 3SH, UK

Prof. G. GABRIELLI, Department of Chemistry, University of Florence, Via G. Capponi 9, I-50121 Firenze

A. GAUDEMER, Université Paris-Sud, UA 255, F-91405 Orsay

Dr R. GRAY, Food and Agricultural Chemistry Research Division, Department of Agriculture for Northern Ireland, Belfast BT9 5PX, Northern Ireland, UK

Prof. A. GUYOT, CNRS - LMO, B.P. 24, F-69390 Vernaison

Dr M.A. HEMMINGA, Laboratory of Dairing and Food Physics, Department of Food Science, Wageningen Agricultural University, PO Box 8128, NL-6700 ET Wageningen

Dr A. JEUNET, LEDSS, Université Joseph Fourier, 15 Rue Maurice Gignoux, F-38031 Grenoble

Dr V. KAMENOPOULOU, Commission à l'Energie Atomique Grecque, G-15310 Aghia Paraskevi Attikis

Dr P. LABRUDE, Université de Nancy 1, France

Dr M. LE MESTE, Laboratoire de Biologie Physico-chimique, Ecole Nationale Supérieure de Biologie Appliquée à la Nutrition et à l'Alimentation, Campus Universitaire Montmuzard, F-21000 Dijon

Dr F. LETERRIER, Unité de Biophysique du C.R.S.S.A., F-38702 La Tronche Cedex

Dr L. LIU, Physicochimie des Rayonnements, Bat. 350, Université Paris Sud UA 75 du CNRS, F-91405 Orsay Cedex

Dr C. LOUIS, Laboratoire de Réactivité de Surface et Structure, URA 1106 CNRS, Université P. et M. Curie, 4 Place Jussieu, F-75252 Paris Cedex 05

Prof. A.L. **MANIERO**, Universita di Padova, Dipartimento di Chimica Fisica, Padova (Italy)

Prof. G. **MARTINI**, Department of Chemistry, University of florence, Via G. Capponi 9, I-50121 Firenze

Prof. M. **MASSOL**, Université Paul Sabatier, Laboratoire de Chimie Bioinorganique, INSERM U-305, 38 Rue des 36 Ponts, F-31400 Toulouse

Dr A. **MICHEL**, Laboratoire des Matériaux Organiques, CNRS, B.P. 34, F-69390 Vernaison

Prof. J.P. **MONTHEARD**, Université de Saint-Etienne, 23 Rue du Dr Michelon;F-42100 Saint-Etienne

Prof. F. **NEPVEU**, Université Paul Sabatier, Laboratoire de Chimie Analytique, 31 Allée Jules Guesde, F-31400 Toulouse

Dr B. **NICKEL-PEPIN-DONAT**, Ecole Normale Supérieure, en détachement au DRFMC/SP2M/PMS, Centre d'Etudes Nucléaires de Grenoble, BP 85X, F-38041 Grenoble Cedex

Prof. M.F. **OTTAVIANI**, Department of Chemistry, University of Florence, Via G. Capponi 9, I-50121 Firenze

Dr R. **PANSU**, Physicochimie des Rayonnements, Bat. 350, Université Paris Sud UA 75 du CNRS, F-91405 Orsay Cedex

Dr G. **PASCAL**, INRA, Laboratoire de Nutrition et Sécurité Alimentaire, F-78352 Jouy-en-Josas Cedex

Dr G. **PORTAL**, Commissariat à l'Energie Atomique, IPSN/DPT/SIDR, B.P. 6, F-92265 Fontenay aux Roses Cedex

Dr G. **POUPEAU**, Institut Dolomieu, Université Joseph Fourier, 15 Rue Maurice Gignoux, F-38031 Grenoble

Dr **David QUERE**, UA 792 du CNRS, Laboratoire de Physique de la Matière Condensée, 11 Place Marcelin Berthelot, F-75231 Paris Cedex 05

Dr J. **RAFFI**, Département de Physiologie du Végétal et Ecosystèmes, CEN Cadarache, F-13108 Saint-Paul-lez-Durance

Prof. A. **RASSAT**, Ecole Normale Supérieure, 24 Rue Lhomond, F-75005 Paris

Dr A. **REVILLON**, Laboratoire des Matériaux Organiques, CNRS, B.P. 34, F-69390 Vernaison

Dr A.M. **RIQUET**, INRA, Laboratoire de Nutrition et Sécurité Alimentaire, F-78352 Jouy-en-Josas Cedex

Dr S. **RIVOMANANA**, Université Paul Sabatier, Laboratoire de Chimie Bioinorganique, INSERM U-305, 38 Rue des 36 Ponts, F-31400 Toulouse

Dr J. **RONCIN**, Physicochimie des Rayonnements, Bat. 350, Université Paris Sud UA 75 du CNRS, F-91405 Orsay Cedex

Dr M.J.G.W. **ROOZEN**, Laboratory of Dairing and Food Physics, Department of Food Science, Wageningen Agricultural University, PO Box 8128, NL-6700 ET Wageningen

Dr A.M. ROSSI, Institut Dolomieu, Université Joseph Fourier, 15 Rue Maurice Gignoux, F-38031 Grenoble

Dr C.C. ROWLANDS, School of Chemistry and Applied Chemistry, University of Wales College of Cardiff, P.O. Box 912, Cardiff CF1 3TB, UK

Y. SANDRAY, Dorlyl, Usine du Havre, B.P. 1152, F-76063 Le Havre Cedex

Prof. U. SEGRE, Universita di Padova, Dipartimento di Chimica Fisica, Padova (Italy)

Dr Z. SOJKA, Jagellonian University, ul. Karasia 3, Krakow, Poland

Dr Edgar SOULIE, Département de Recherches sur l'Etat Condensé, DRECAM/SCM/BP 121, Centre D'Etudes Nucléaires de Saclay, F-91191 Gif sur Yvette Cedex

Prof. M. H. STEVENSON, Food and Agricultural Chemistry Research Division, Department of Agriculture for Northern Ireland, The Queen's University of Belfast, Newforge Lane, Belfast BT9 5PX, Northern Ireland, UK

Dr C.L. THIERY, Laboratoire de Radiochimie des Constituants des Aliments, Département de Physiologie Végétale et Ecosystèmes, CEN de Cadarache, F-13108 Saint-Paul-lez-Durance

Dr J.M. THIERY, Modélisation et Génie Logiciel, Département de Physiologie Végétale et Ecosystèmes, CEN de Cadarache, F-13108 Saint-Paul-lez-Durance

Dr B. TILQUIN, Laboratoire d'analyse chimique et physico-chimique des médicaments, CHAM 72.30, 72 Avenue E. Mounier, B-1200 Bruxelles

Dr J.B. TRIOLET, Laboratoire de Radiochimie des Constituants des Aliments, Département de Physiologie Végétale et Ecosystèmes, CEN de Cadarache, F-13108 Saint-Paul-lez-Durance

Dr T. VAN LIET, Laboratory of Dairing and Food Physics, Department of Food Science, Wageningen Agricultural University, PO Box 8128, NL-6700 ET Wageningen

Dr V. VERNEY, Laboratoire des Matériaux Organiques, CNRS, B.P. 34, F-69390 Vernaison

Dr P.B. VINCENT, Laboratoire de Radiochimie des Constituants des Aliments, Département de Physiologie Végétale et Ecosystèmes, CEN de Cadarache, F-13108 Saint-Paul-lez-Durance

Dr J. VIRET, Unité de Biophysique du C.R.S.S.A., F-38702 La Tronche Cedex

Dr P. WALSTRA, Laboratory of Dairing and Food Physics, Department of Food Science, Wageningen Agricultural University, PO Box 8128, NL-6700 ET Wageningen

Dr C. WOLF, URA CNRS 524, C H U St-Antoine, 27 Rue Chaligny, F-75012 Paris

Dr F. ZEEGERS, Laboratoire d'analyse chimique et physico-chimique des médicaments, CHAM 72.30, 72 Avenue E. Mounier, B-1200 Bruxelles

Dr J.L. ZIMMERMANN, Service de Biophysique, Département de Biologie, CEN de Saclay, F-91191 Gif sur Yvette

Contents

Some Recent Applications of Computer Simulations of ESR Spectra to the Dynamics of Spin Probes in Gels Membranes

Claude CHACHATY, Edgar SOULIE
Département de Recherches sur l'Etat Condensé
DRECAM/SCM/BP 121, Centre d'Etudes Nucléaires de Saclay.
91191 Gif-sur-Yvette cedex, France.

Claude WOLF
URA CNRS 524, CHU St. Antoine, 27 rue Chaligny
75012 Paris.

ABSTRACT

The ESR of S=1/2 paramagnetic species is one of the most convenient methods for probing microheterogeneous structures in gels, lyotropic liquid crystals or membranes. The interpretation of ESR spectra in term of dynamics of the spin probes generally requires computer simulations whose principles are briefly described. Some examples are given of simulations applied to the determination of the flexibility of the tridimensional network in polymer gels crosslinked by Cu^{2+} and VO^{2+} as well as to the dynamics of nitroxide spin probes in gels and in phospholipid model membranes.

RESUME

La RPE de sondes paramagnétiques de spin S=1/2 est l'une des methodes les mieux appropriées à l'étude des structures microhétérogènes dans les gels, les cristaux liquides lyotropes et les membranes. L'interprétation des spectres de RPE en termes de dynamique nécessite généralement des simulations sur ordinateur dont on décrit brièvement les principes. On donne quelques exemples d'application de ces simulations à la détemination de la flexibilité du réseau tridimensionnel de gels de polymères réticulés par Cu^{2+} et VO^{2+} ainsi qu'à la dynamique de sondes nitroxydes dans les gels et dans les membranes phospholipidiques modèles.

B. Catoire (Editor)
Electron Spin Resonance (ESR) Applications in Organic and Bioorganic Materials

INTRODUCTION

The applications of ESR to the dynamics of S=1/2 paramagnetic probes have been initiated by the works of McConnell[1] and Kivelson[2]. They were initially restricted to the case of fluid solutions where the reorientational motions are fast enough for averaging to zero the anisotropies of the hyperfine coupling **A** and spectroscopic **g** tensors. Under these conditions the homogeneous line width is inversely proportional to the electron spin relaxation time T_2. The peak to peak linewidth ΔH_{pp} of the first derivative is given by a relation of the form[3] :

$$\Delta H_{pp} = \frac{2}{\gamma T_2 \sqrt{3}} + \Delta H_R = a + bm_I + cm_I^2 + \Delta H_R \qquad [1]$$

where m_I is the nuclear quantum number, a, b and c being constants depending on the principal values of the **A** and **g** tensors as well as on the reorientation correlation times. ΔH_R is the inhomogeneous contribution to the linewidth resulting for instance from unresolved hyperfine splittings[4].

In the case of macromolecules, micellar solutions and more generally of heterogeneous systems, the molecular reorientation is often too slow to allow a straightforward determination of the reorientation correlation times by measurements of the dependence of ΔH_{pp} upon m_I. The position, shape and width of the lines then become dependent on these correlation times. In isotropic media, the transition between the fast and slow motional regimes is evidenced by an increase of the total width of the spectrum with the occurence of side bands whose separation provides an estimate of the motional correlation times. For axially symmetric **A** and **g** tensors, the condition of slow motional regime corresponds to :

$$\tau^{-1} \lesssim (\gamma_{/\!/} - \gamma_{\perp})B_0 + \gamma_0 (A_{/\!/} - A_{\perp}) \qquad [2]$$

where τ is an effective motional correlation time, $\gamma_0 = (g_{//} + 2g_{\perp})\beta/\hbar$ the electron magnetogyric ratio and B_0 the magnetic field of the spectrometer. In X band (9 - 10 Ghz) the transition between the two regimes corresponds approximately to τ = 0.2, 0.4 and 2 ns, in the case of electron spin ½ probes like Cu^{2+}, VO^{2+} and nitroxide radicals, respectively.

THEORETICAL BACKGROUND

A number of works have been devoted to the simulation of ESR spectra in the domain of slow motions. The most important contribution to this problem has been brought on since the early seventies by Freed and his coworkers[5-10]. Several simplified treatments have been proposed by Freed[8], Kivelson[11] and more recently by Moro[12]. For the study of molecular motions in heterogeneous systems, in particular gels and resins of crosslinked polymers, we have written several computer programs which may be applied to the ESR of spin S=1/2 probes as well as to the NMR of nuclei presenting a strong chemical shift anisotropy such as ^{31}P or quadrupolar nuclei like ^{2}H and ^{17}O. They run in interactive mode and are written in APL language. A Fortran version with an APL interface also exists for one of them. These programs first compute the absorption spectrum which may be convoluted to take into account a gaussian or lorentzian broadening independent of the motion. The ESR spectra are then numerically derived for comparison with the experimental ones. In these programs we have considered two basic models of reorientation :

The finite jump model where the tumbling occurs randomly among any orientation with respect to the magnetic field.The adustable parameters of the simulation are the lifetime τ of an orientation and the intrinsic line widths. The line shape is given by the imaginary (absorption) or real (dispersion) parts of the expression:

$$G(B_0) = i\gamma B_1 M_0^m \frac{\sum_j P_j \left(T_{2j}^{-1} + \tau^{-1} + i\,\gamma\,(B_j - B_0)\right)^{-1}}{1-\tau^{-1} \sum_j P_j \left(T_{2j}^{-1} + \tau^{-1} + i\,\gamma\,(B_j - B_0)\right)^{-1}} \qquad [3]$$

where M_0^m is the longitudinal magnetization for the nuclear quantum number m, $P_j \propto \sin\theta_j$ the probability of the orientation j, B_j the corresponding resonance field and B_1 the r.f. field. The above expression is derived from Bloch equations modified for multisite exchange[13]. We have applied this model to the simulation of the ^{31}P NMR spectra in liquid crystals[14] as well as to the ESR of a gel of copper complex[15]. In the latter case, the correlation time τ was derived in a simple way from the ratio of the **g** tensor anisotropies measured at 9 and at 35 Ghz.

The small step model (or Brownian diffusion) where the reorientation occurs by infinitesimal rotations. The simulated spectrum is obtained by solving for $\hat{M}^m_j$ a system of coupled Bloch equations modified for diffusion :

$$i\gamma B_1 M_0^m P_j = \hat{M}^m_j \left(i\gamma (B_j - B_0) + T_{2j}^{-1} + k_j^- \tau^{-1} \right)$$

$$-\tau^{-1} \left(k^+_{j-1} \hat{M}^m_{j-1} + k^+_{j+1} \hat{M}^m_{j+1} \right) \qquad [4]$$

valid under slow passage conditions for a non saturating r.f. field. Here the reorientation correlation time is $\tau = (6D)^{-1}$, D being the rotational diffusion coefficient. $\hat{M}^m_j = u^m_j + iv^m_j$ is the transverse magnetization for the orientation j. The rate of magnetization transfer from an infinitesimal angular zone centered at θ_j to the adjacent ones is $k^-_j \tau^{-1}$, that of the reverse transfer being $k^+_{j\pm1} \tau^{-1}$. The expressions of the coefficients k^-_j and $k^+_{j\pm1}$ are given by McCalley et al.[16] The other parameters have the same meanings as in equ. [3]. In equations [3] and [4] the angular dependence of T_2, i.e. of the intrinsic linewidth, is assumed to have the form :

$$T_{2j}^{-1} = \left(\Delta\omega^2_{/\!/} \cos^2 \theta_j + \Delta\omega^2_{\perp} \sin^2 \theta_j \right)^{1/2} \qquad [5]$$

$\Delta\omega_{/\!/}$ and $\Delta\omega_{\perp}$ being the half-widths at half-heigth of absorption lines expressed in radians s^{-1}.

These equations hold strictly for isotropic reorientations and axially symmetric tensors. They may be extended with some caution to more general cases. For a given value of the correlation time τ, the two basic models considered here give appreciably different spectral shapes : the line broadening is larger for the jump model than for the small step one, whereas the reverse is observed for the line shifts.

APPLICATIONS

1) Polymer gels

We are presently working on problems of molecular diffusion in polymer gels. The model system chosen for this study is the poly-4-vinylpyridine (P4VP) crosslinked by divalent ions such as Cd^{2+}, VO^{2+} and Cu^{2+} (17). The main techniques involved in this work are NMR and nuclear relaxation, ESR and small angle X-ray scattering. In complement to NMR, ESR provides information on the flexibility of the tridimensional network and the time scale of exchange between the pores or cavities existing in the gels.

The mobility of divalent ions at the nodes of the network has been obtained from the ESR of VO^{2+} or Cu^{2+}, using spectral simulations where the main parameter is the correlation time τ_R for an isotropic brownian motion . For the vanadyl ions, these simulations have been performed on selected spectra recorded in the 77-350 K range (Figure 1). As this procedure is somewhat tedious and time consuming, we have derived from these simulations empirical relationships allowing a simple estimate of τ_R from the shifts of some spectral features.

It has been shown for nitroxide radicals(8) as well as for VO^{2+} (18) that in the slow motional regime, the correlation time may be obtained from a relation of the form:

$$\tau_R = a(1 - S_1)^{-b} \quad [6]$$

where a and b are constants dependent on the reorientation model and of the intrinsic line width. S_1 is the ratio of the distance Δ_1 between the two extreme parallel edge lines of the spectrum to its rigid limit equal to $2IA_{zz}$, I being the spin of the nucleus giving the main hyperfine structure and A_{zz} the largest component of the **A** tensor. In the case of the P4VP-VO^{2+} gel, we have observed that this method is limited to the 2-10 ns range either by an excessive line broadening or by a very small shift of the parallel edge lines ($m_I = \pm 7/2$) at the lowest and highest limits of this range, respectively. Our computer simulations have shown that equ. [6] also holds for the separation Δ_2 between the $m_I = -5/2$ and $m_I = +5/2$ parallel edge lines with $S_2 = \Delta_2 / 5A_{zz}$ as well as for the separation Δ_3 between a parallel ($m_I = -5/2$) and a perpendicular ($m_I = -7/2$) singularity (Figure 2). The measurement of Δ_3 allows to extend the determination of τ_R up to 100 ns. The

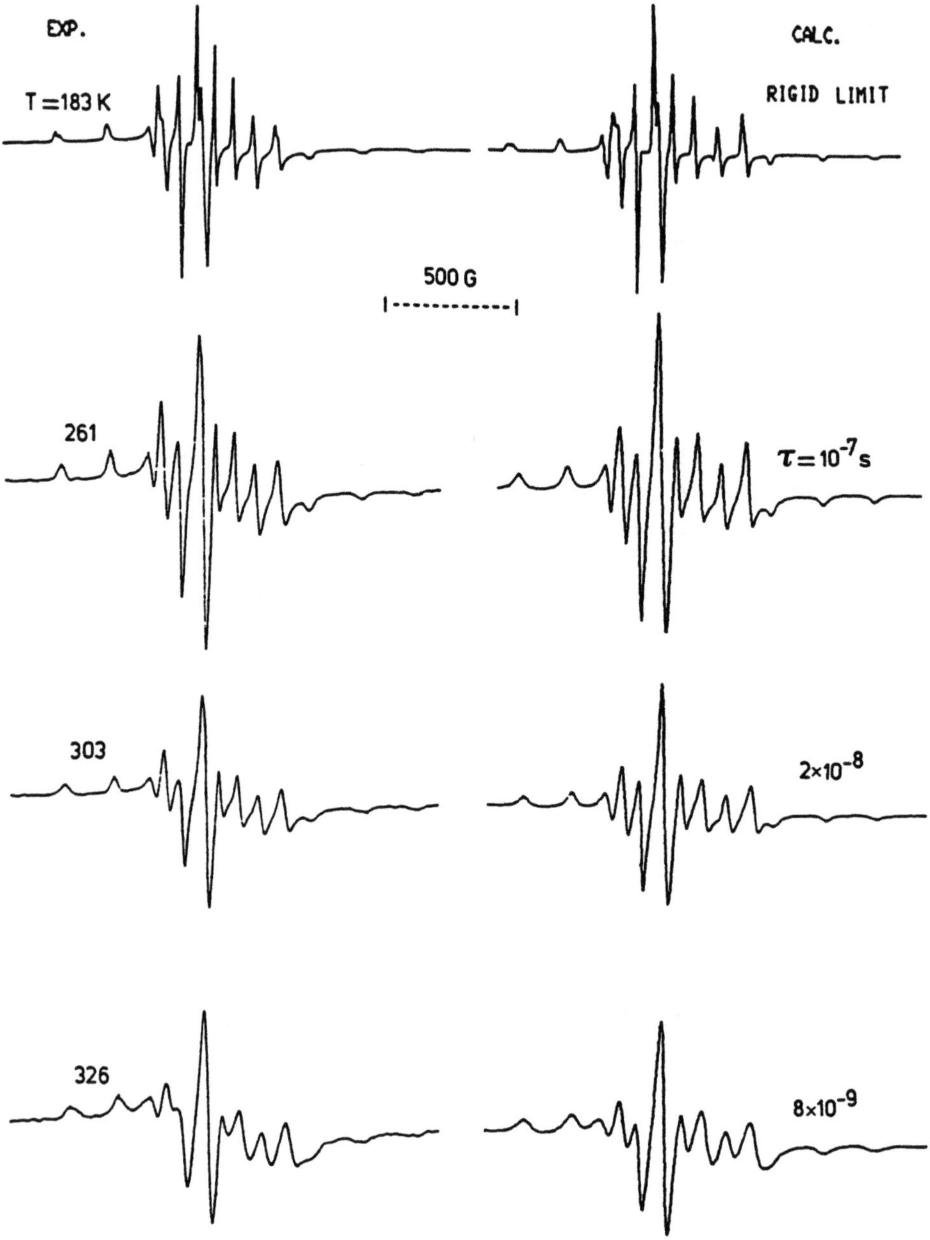

Figure 1

Computer simulations of the vanadyl spectra of a P4VP-VO^{2+} swollen by ethanol, taking $A_{/\!/}$ = 190, $A_{\perp}$ = 68 gauss, $g_{/\!/}$ = 1.945, $g_{\perp}$ = 1.980. The line width are $\Delta H_{/\!/}$ = 20 G and $\Delta H_{\perp}$ = 15 G.

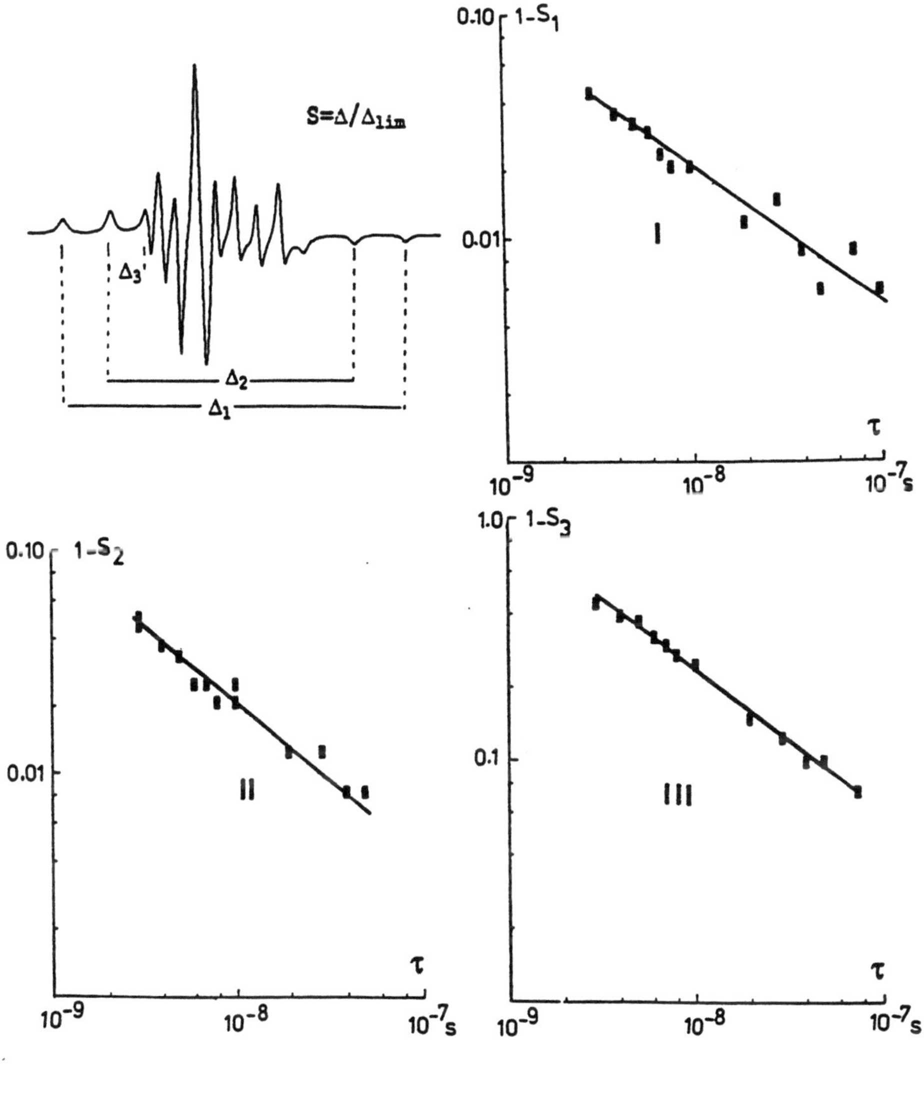

Figure 2

Least square determination of the parameters a and b of equ. [6] for the estimate of the reorientation correlation time τ_R from the ratios Δ_1, Δ_2 and Δ_3 splittings of the VO^{2+} spectrum to their rigid limit . These parameters are :

(I) a = 1.653x10^{-11} s, b = 1.667 for Δ_1.

(II) a = 3.693x10^{-11} s, b = 1.451 "" Δ_2.

(III) a = 8.312x10^{-10} s, b = 1.734 "" Δ_3.

consistency between the τ_R values derived from Δ_1, Δ_2 and Δ_3 has been verified by an Arrhenius plot covering the 180 - 330 K range. Figure 4 shows the influence of the swelling by ethanol on the rigidity of the P4VP-VO^{2+} gel.

A similar analysis has been performed on P4VP gels crosslinked by Cu^{2+}. In this case the line shifts can only be accurately measured on the second derivative of the absorption spectrum (Figure 3). The superhyperfine structure observed under these conditions corresponds to the coupling of the unpaired electron spin with two equivalent nitrogens giving rise to a well resolved 1-2-3-2-1 quintet. This confirms that on the average two pyridine groups are involved in the interchain or intrachain crosslinks.

The temperature dependence of the spectra of P4VP-Cu^{2+} gels is governed by the anisotropy Δg of the **g** tensor. Computer simulations of the Cu^{2+} spectra with $g_{zz} = 2.264$, $g_{xx} = g_{yy} = 2.068$ and $A_{zz} = 174$ G, $A_{xx} = A_{yy} = 45$ G show that the τ_R dependence of the reduced **g** anisotropy $\Delta g/(g_{zz} - 0.5(g_{xx} + g_{yy}))$ is also given by equ.[6], the a and b parameters being strongly dependent on the magnetic field strength.

For the same metal/P4VP ratio, the temperature dependence of τ_R is about the same as for VO^{2+} showing that the nature of the divalent cation ensuring the crosslinking has virtually no influence on the rigidity of the gel network (Figure 4). This is a useful information for the interpretation of nuclear relaxation in such systems which depend on the fast segmental motions of the polymer chains as well as on the deformation of the gel network occuring at a much longer timescale(19).

We have also investigated the dynamical behaviour of a small probe, the di-tertiobutyl nitroxide (DTBN) in a P4VP gel crosslinked by Cd^{2+} and swollen with ethanol. Below 230 K the DTBN spectrum corresponds to a very slow motion ($\tau_R \gtrsim 10$ ns). At this temperature appears a narrow component corresponding to a fast motion ($\tau_R \lesssim 0.1$ ns) . As the temperature is raised, these two spectral components progressively collapse up to 270 K where a spectrum corresponding to a fast anisotropic motion is obtained (Figures 5 and 6). This behaviour has been assigned to an exchange of the DTBN probe among sites of different reorientational freedom probably related to a distribution of pore sizes in the gel. For simplicity, we assume that the exchange occurs among two sites corresponding to the fast and slow motional regimes as defined by

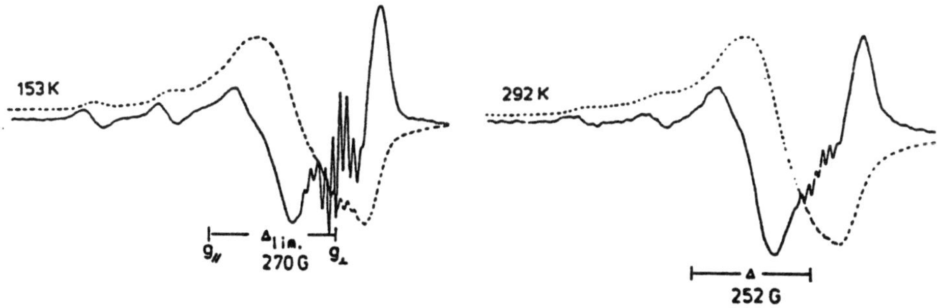

Figure 3

First (-----) and second derivatives of absorption spectra of a P4VP-Cu^{2+} gel swollen by ethanol. Computer simulations not given here show that the dependence of the **g** tensor anisotropy upon τ_R is also given by equ. [6] with a = 3.34x10^{-10} s and b = 0.98.

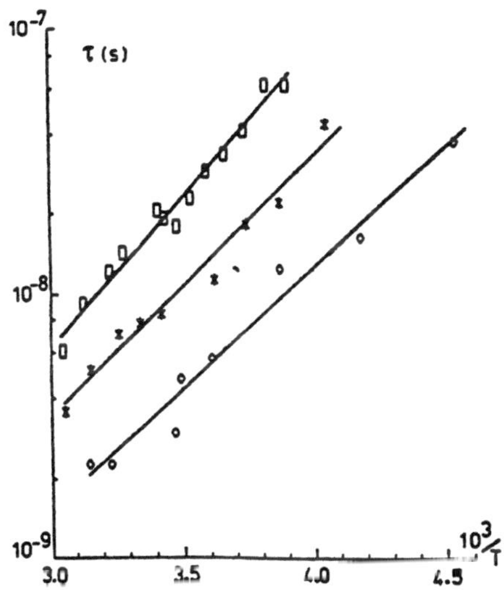

Figure 4

Arrhenius plot of the motional correlation time of divalent ions at the network nodes of P4VP gels swollen by ethanol. Effect of the swelling ratio ρ.

The experimental data have been fitted using the expression $\tau_R = \tau_0 \exp(E/RT)$ with :

(□) τ_0 = 2.2x10^{-12} s, E = 21.9 kJ/mole (P4VP-VO^{2+},ρ = 2.5)

(*) τ_0 = 3.3x10^{-12} s, E = 19.1 kJ/mole (P4VP-Cu^{2+}, ρ = 6)

(o) τ_0 = 2.55x10^{-12}s, E = 17.6 kJ/mole (P4VP-VO^{2+}, ρ = 12)

For these three samples, M^{2+}/P4VP ≃ 1/100.

equ. [2]. The absorption spectrum correspond to the imaginary part of the complex transverse magnetization :

$$\hat{M} = f\,\hat{M}_F + (1 - f)\,\hat{M}_S \qquad [7]$$

f being the fraction of probes undergoing a fast reorientation. For each value of B_0 , $\hat{M}$ is obtained by solving a system of N+1 Bloch equations , N being the number of angular zones defined above[20] :

$$i\gamma B_1 M^m_{S_j} = \hat{M}^m_{S_j}\left(i\gamma\left(B^m_{S_j} - B_0\right) + T^{-1}_{2S_j} + k^-_j\tau_S^{-1} + K_{S_jF}\right)$$

$$- \tau_S^{-1}\left(k^+_{j-1}\hat{M}^m_{S_{j-1}} + k^+_{j+1}\hat{M}^m_{S_{j+1}}\right) - K_{FS_j}\hat{M}^m_F \qquad [8a]$$

$$i\gamma B_1 M^m_F = \hat{M}^m_F\left(i\gamma\left(B^m_F - B_0\right) + \left(T^m_{2F}\right)^{-1} + \sum_{j=1}^{N} K_{FS_j}\right)$$

$$- \sum_{j=1}^{N} K_{S_jF}\,\hat{M}^m_{S_j} \qquad [8b]$$

where the subscripts S and F refer to the slow and fast reorientation sites. The j subscript has the same meaning as in equ. [4]. The average exchange rate between these sites is :

$$\nu_{ex} = fK_{FS} = (1 - f)K_{SF} \qquad [9]$$

with $K_{SF} = K_{S_jF}$ and $K_{FS} = K_{FS_j}/P_j$. The spectra of Figure 5 have been simulated taking as adjustable parameters the population f of F sites, the exchange rate ν_{ex} as well as the reorientation correlation times at these two sites . For site S, a single correlation time was estimated from the total width of the spectrum using a calibration curve derived from computer simulations. For site F, two correlation times were used to calculate T^m_{2F} by means of equ. [1] whose coefficients a,b and c are given in ref.(21) for an anisotropic motion. In the 230 - 270 K range, the exchange rate increases from 10^5 to 10^7 s^{-1}. At higher temperatures, the DTBN probe undergoes a fast anisotropic motion about an axis perpendicular to the directions of the N-O bond and of the $2p_z$ unpaired spin orbital. These ESR experiments on DTBN are helpful for a more general study in progress on the dynamical behaviour of solvents in gels using nuclear relaxation.

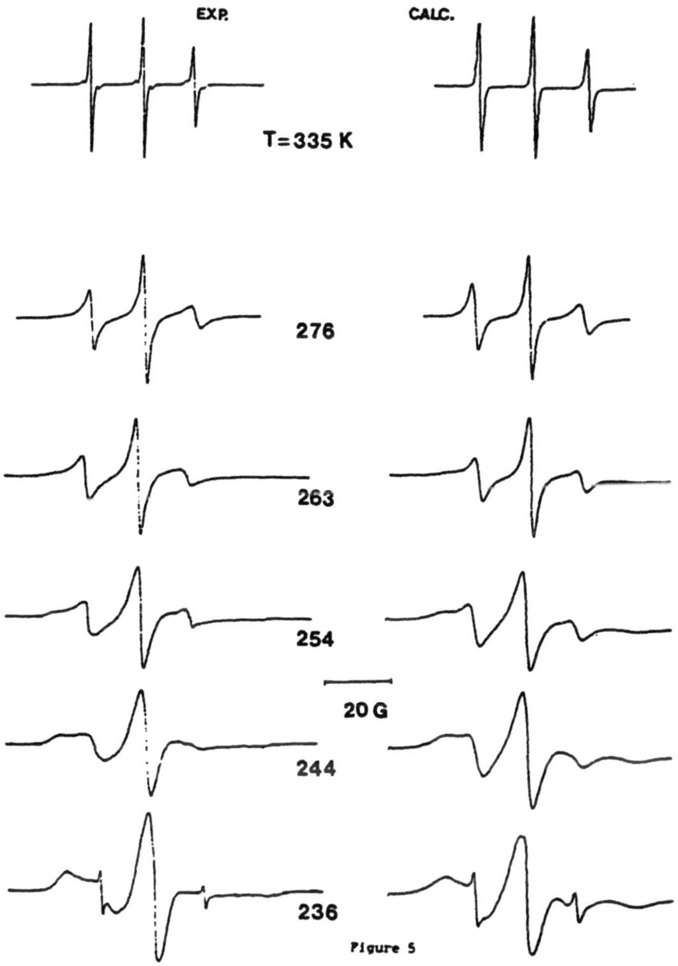

Figure 5.

Spectra of di-tertiobutyl nitroxide in a P4VP-Cd^{2+} gel swollen by ethanol. Below 0°C the spectra have been simulated by means of equ. [8] (slow or intermediate multisite exchange), taking g_{zz} = 2.0027, g_{yy} = 2.0062, g_{xx} = 2.0087 and A_{zz} = 33.7 G, A_{yy} = 6.3 G, A_{xx} = 8.05 G. Above this temperature, the exchange is fast and the correlation times $\tau_{/\!/}$ and $\tau_{\perp}$ are obtained by a least squares adjustment of line widths. The main parameters for these simulations are as follows :

T = 236 K, f = 0.015, ν_{ex} = 10^5 s^{-1}, τ_s = 10 ns, $\tau_{f/\!/}$ = 0.1 ns, $\tau_{f\perp}$ = 0.3 ns
T = 244 K, f = 0.05, ν_{ex} = 3×10^6 s^{-1}, " " "
T = 254 K, f = 0.10, ν_{ex} = 5×10^6 s^{-1}, " " "
T = 263 K, f = 0.20, ν_{ex} = 7.5×10^6 s^{-1}, " " "
T = 276 K (fast exchange), $\tau_{/\!/}$ = 0.38 ns, $\tau_{\perp}$ = 2.6 ns.
T = 335 K " " ", $\tau_{/\!/}$ = 0.24 ns, $\tau_{\perp}$ = 0.39 ns.

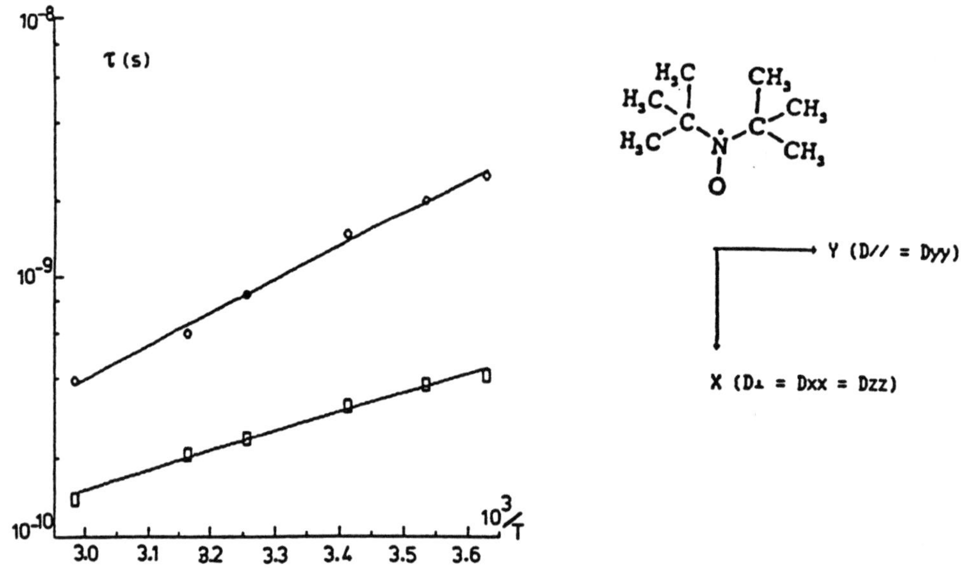

Figure 6.
Arrhenius plots of the reorientation correlation times of TBN in a P4VP-Cd^{2+} gel fitted with $\tau = \tau_0 \exp(E/RT)$, taking $\tau_0 = 5.54 \times 10^{-14}$, E = 24.5 kJ/mole and $\tau_0 = 8.9 \times 10^{-13}$ s, E = 14.2 kJ for $\tau_\perp$ (o) and $\tau_{//}$ (□), respectively.

2) Lyotropic liquid crystals and biological membranes.

In the case of organizéd assemblies of amphiphilic molecules forming for instance lamellar or hexagonal liquid crystalline phases, the anisotropy of a magnetic tensor such as **A** or **g** is not averaged out to zero even by a fast reorientational motion. The substrate molecules or paramagnetic probes indeed undergo a restricted librational motion about the director (optical axis) of the mesophase, which is often depicted by the simplest model of wobbling in a cone (Figure 7). In the case of a randomly oriented sample, one can define a local director Δ_L which corresponds to the direction perpendicular to the bilayers (lamellar phase) or to the axis of the cylinders (hexagonal phase). The magnetic resonance spectra then are somewhat similar to solid state spectra and depend

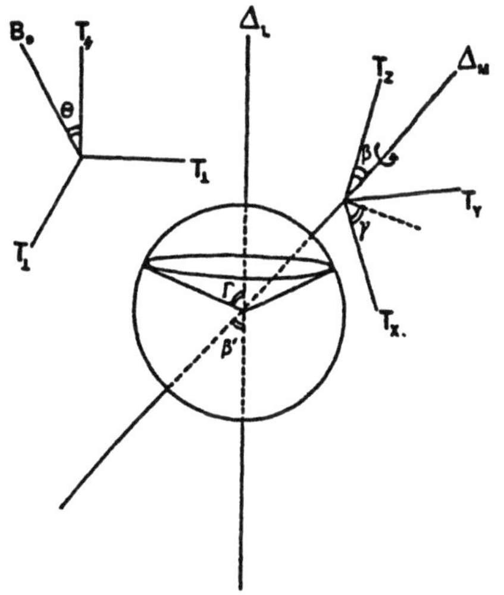

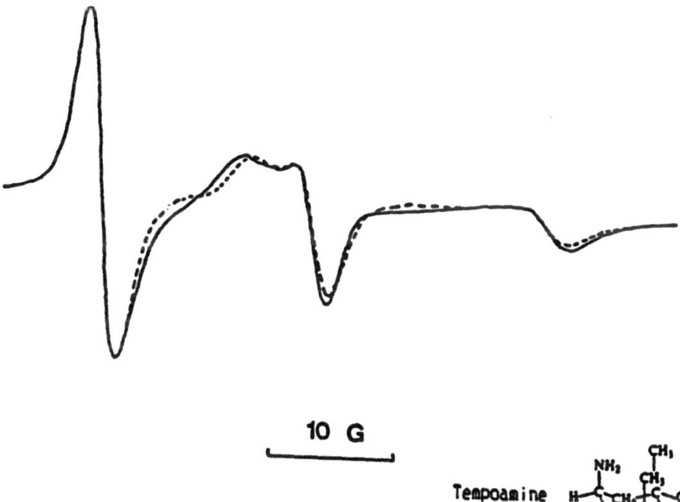

Tempoamine 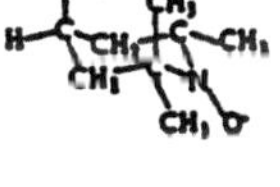

Figure 7.

Top : Definition of angles and axes involved in the ESR of a spin probe dissolved in a uniaxial liquid crystal. Δ_L is the local director, Δ_M the long axis of the molecular diffusion tensor. T_x, T_y and T_z are the principal values of the **A** or **g** tensors which are averaged to $T_{/\!/}$ and $T_\perp$ by reorientation about Δ_M and Δ_L.
Bottom : spectrum of tempoamine in randomly oriented sodium dibutyl-phosphate/water lamellar phase[22] .The dashed spectrum has been calculated with $\tau_{/\!/} = 0.5$ ns, $\tau_\perp = 2.5$ ns, assuming that the long axis of the diffusion tensor **D** is parallel to the N-O bond (X axis). The molecular order parameter $\overline{P}_2(\cos \beta')$ is 0.7 .

upon the orientation of the tensors with respect to B_0. If the molecular tumbling motion is fast enough, the tensors are averaged out and become axially symmetric about Δ_L. In this case the splittings and shifts of the lines with respect to the isotropic liquid provide the order parameters, i.e. the eigenvalues of the tensor :

$$S_{\alpha\beta} = \frac{1}{2} < 3\, l_\alpha l_\beta - \delta_{\alpha\beta} > \qquad [10]$$

where l_α and l_β are the direction cosines of Δ_L in the x,y,z molecular frame (Figure 7). The order parameter corresponding to the largest absolute eigenvalue of the ordering tensor is often denoted as S_{33}. If the motion of the probe is axially or quasi-axially symmetric about a molecular axis Δ_M one can define a molecular order parameter as :

$$S_{33} \cong \overline{P}_2 (\cos \beta') = \frac{1}{2} <3 \cos \beta' - 1 > \qquad [11]$$

where β' is the time dependent angle between Δ_M and Δ_L. The asymmetry parameter $(S_{11} - S_{22})/S_{33}$ may be estimated from the residual **A** and **g** anisotropies.

A program has been written in APL to simulate the spectra of probes undergoing a fast tumbling motion in liquid crystalline phases[22,23]. In this case the line widths are given by an expression of the same form as equ. [1] where the a,b and c coefficients depend on the same parameters as in isotropic liquids but also on the order parameters and on the angle Θ between Δ_L and B_0. This program allows the simulation of spectra in monooriented or randomly oriented samples (Figure 7). The degree of orientation is conveniently represented by the standard deviation $\Delta\Theta$ of a Gaussian angular distribution about a mean value Θ_0.

An interesting application of this program is the study of the orientational and dynamical behaviour of nitroxide spin probes in biological membranes. It has been recently applied to works on the interaction of relevant biological agonists (asymmetric phospholipids such as PAF-acether[24]) and lipophilic drugs (e.g. antiinflammatory corticosteroids or neuroleptic phenothiazines[20]) with model membranes.

Fully hydrated multibilayers of egg lecithin are relatively reliable models of simple biomembranes. They can be easily prepared from a biological compound with a biologically significant variety of acyl chains. They constitute stacks of bilayers which can be

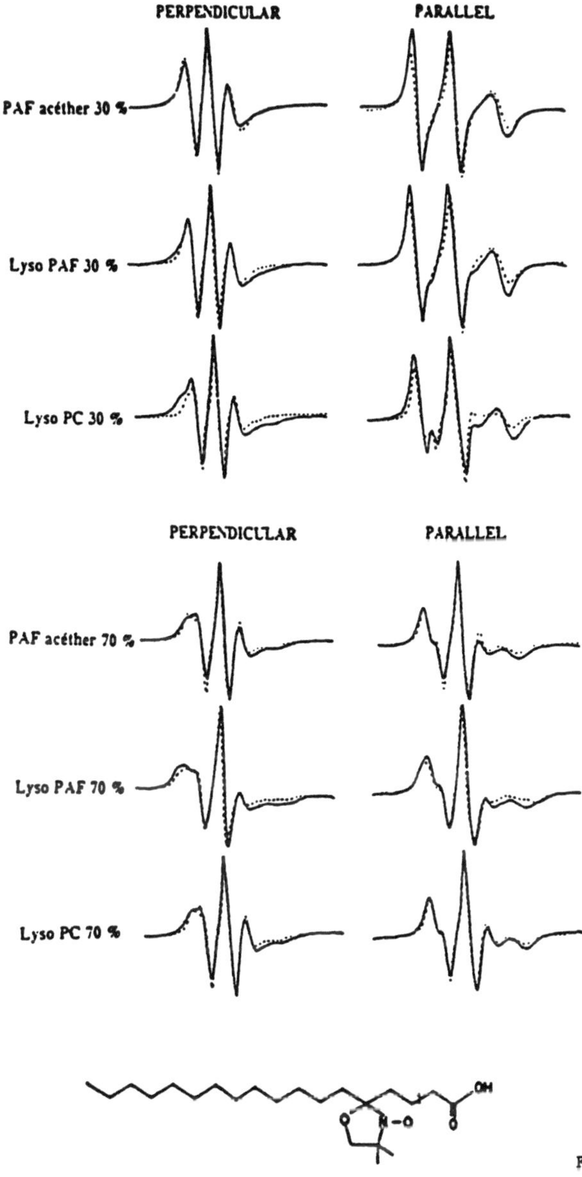

5 NS

Figure 8

Figure 8.

Spectra of 5 doxylstearic acid (5NS) intercalated in oriented multibilayers of egg lecithin containing different asymmetric phospholipids in proportions indicated on each spectra [24]. The simulated spectra (dashed lines) were obtained using numerical values of $\Delta\Theta$ and S_{33} given in Figure 9 and taking τ_R = 1.5 ns. In the case of 70% lyso PAF in egg lecithin, the spectrum was simulated as a superimposition of two poorly oriented components with S_{33}=0.5 (60%) and 0.85 (40%).

oriented by spreading over a glass slide and then rotated with respect to the magnetic field. When probed by a nitroxide labelled fatty acid or by a spin labelled steroid, which are self-orienting probes inside the bilayers, the liquid crystalline nature of the stacked membranes is evidenced by the strong anisotropy of the ESR spectra. Spectral simulations offer a quantitative approach to the parameters characterizing the liquid crystalline membrane such as the orientation distribution of the local directors Δ_L (macroscopic order correlated with the long range order of the phase as evidenced by X-ray diffraction), molecular ordering ($\overline{P}_2(\cos \beta')$) and viscodynamic properties probed by the rotational correlation times. The amount of information provided by the ESR of spin probes is unique to biologists and compares favourably with other improved techniques of membranology. The ESR however entails some specific difficulties related for instance to the incorporation of spin probes into the matrix and to the necessity of a constant hydration degree during the experiments. In the studies reported here, the films are constituted by one of the major components of natural membranes (phosphatidylethanolamine, acidic phospholipids, sphyngolipids...) which can also form hexagonal II and cubic phases. Although the transition between the lamellar (multibilayers) phase and one of these phases may be detected by the ESR of spin probes, their identity has to be established by X-ray diffraction.

As an example of such studies, we have chosen a recent work on the progressive disorientation of phosphatidylcholine multibilayers upon addition of an asymmetric phospholipid such as PAF-acether[24]. This compound, a potent activation factor in many cell types, considerably broadens the orientation distribution of the director of the lamellar phase whereas the order parameter of the probe is slightly affected. The quantitative aspects of these data can only be obtained by simulation of the spin probe spectra (Figures 8 and 9).

The ESR data are correlated with the appearance of an interdigitated lamellar phase evidenced by X-ray diffraction . They are also in agreement with ^{31}P NMR experiments which display a broadened signal for oriented multilayers of lecithin upon addition of PAF-acether 40%. The results of ESR, NMR and X-ray experiments led us to propose a model of "blisterized" membrane (Figure 10). This model presenting a dramatically altered membrane stucture is

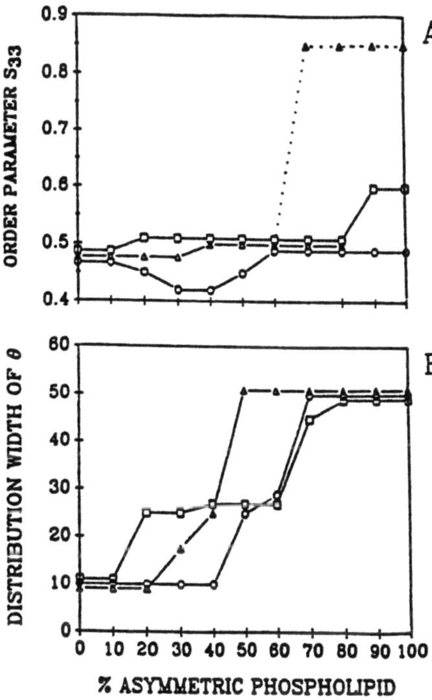

Figure 9.
Order parameters (A) and standard deviation ΔΘ of the orientation distribution of the local director (B) for 5NS probing oriented egg lecithin multibilayers.
-Δ- Lyso-PAF.
-▲- Lyso-PAF (second component).
-o- PAF-acether.
-□- Lysophosphatidylcholine.

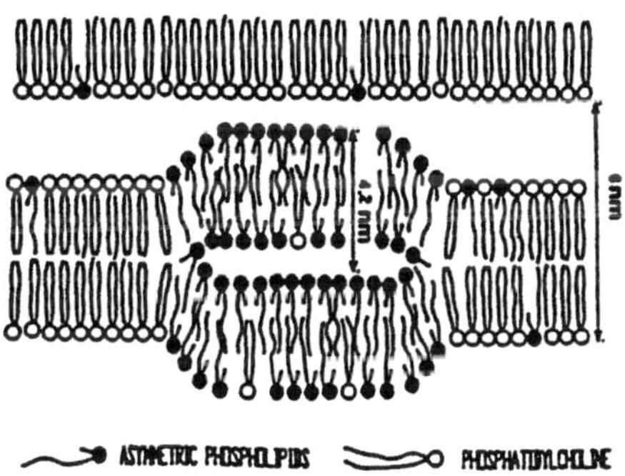

Figure 10
Model for the structure of PAF-acether/PC films.

in agreement with the physiological event following cell exposure to PAF-acether. The discrepancy between the assumptions based on the ESR, NMR and X-ray data and the biological observation arises from the strong difference in the concentrations required to physically alterate the membrane (over 35% PAF-acether) and to induce cell activation (lower than 10^{-10} molar for Guinea pig platelets). However this can be explained by a large difference between a weak average concentration and a strong local concentration around a quasi-receptor specific of PAF.

The binding of several spin labelled phenothiazines to intact red cells[20] is also an appropriate example of ESR lineshape simulation as a tool for obtaining quantitative information on drug-biomembrane interactions. Here the positively charged amphiphilic neuroleptic drug is bound to two types of sites inside the membrane, one where it tumbles slowly and isotropically and one where it undergoes a fast anisotropic motion. The temperature dependence of the exchange among these two sites has been investigated by simulating the spectra using equations [8a] and [8b].From a comparison between different derivatives of the drug (active or not) one concludes that the existence of a positive charge on the side-chain of the probe is essential for binding to a protein membrane where it undergoes a slow reorientational motion whereas it reorients rapidly in lipidic domains independently of its charge.

CONCLUSION

It is worthwhile to compare the application of ESR and NMR to the physical-chemistry of macromolecular systems or organized molecular aggregates (micelles, liquid crystals) and more generally of microheterogeneous systems.

These two techniques involve comparable difficulties of interpretation often requiring sophisticated computer programs, in particular for spectral simulations. T_1 and T_2 relaxation times measurements,providing the most accurate information on the dynamics of substrate molecules or probes, are routinely performed in NMR but not in ESR because up to now, the pulsed spectrometers are not of wide-spread use. Another weak point of the ESR is the lack of specificity,so that the support of other techniques is generally required. On the other hand, the ESR of spin probes is intrinsically more sensitive than NMR and the electron spin relaxation times are much shorter than the nuclear ones, allowing a rapid exploration of the influence of experimental parameters such as temperature or dilution on the physico-chemical properties of a system.

REFERENCES

1 - McConnell H.M., J. Chem. Phys. **25**, 709 (1956).

2 - Kivelson D., J. Chem. Phys. **27**, 1087 (1957).

3 - Kivelson D., J. Chem. Phys. **33**, 1094 (1960).

4 - Bales B.L., J. Magn. Res. **38**, 193 (1980).

5- Freed J. H., Bruno G.V., Polnaszek, J. Phys. Chem.**75**, 3385 (1971).

6- Goldman S.A., Bruno G.V., Polnaszek C.F., Freed J.H., J. Chem. Phys., **56**, 716 (1972).

7 - Goldman S.A., Bruno G.V., J.H. Freed, J. Phys. Chem. **76**, 1858 (1972).

8 - Freed J.H. in "Spin Labeling, Theory and Applications" Chap. 3, Academic Press, N.Y. 1976.

9 - Campbell R.F., Freed J.H., J. Chem. Phys. **84**, 2668 (1980).

10- Schneider D.J., Freed J.H. in "Biological Magnetic Resonance 8. Spin Labeling Theory and Applications" chap.1,edited by L.J. Berliner and J. Reuben, Plenum Press N.Y. (1989).

11- Kivelson D., Lee S., J. Chem. Phys., **76**, 5746 (1982).

12- Moro G., Segre U., J. Magn. Res. **83**, 65 (1989).

13- McConnell H.M., J. Chem. Phys. **28**, 430 (1958).

14- Chachaty C., Quaegebeur J.-P., Caniparoli J.-Ph., Korb J.-P., J. Phys. Chem., **90**, 1115 (1986).

15- Terech P., Chachaty C., Gaillard J., Giroud-Godquin A.M., J. Physique, **48**, 663 (1987).

16- McCalley R.C., Shimshick F.J., McConnell H.M., Chem. Phys. Letters, **13**, 115 (1972).

17- Korb J.-P., Sapoval B., Chachaty C., Tistchenko A.M. , J. Phys. Chem., **94**, 953 (1990).

18- Bruno G.V., Eastman M.P., Harring J.K., J. Phys. Chem., **30**, 1892 (1976).

19- Brereton M.G., Macromolecules **23**, 1115 (1990).

20- Olivier J.-L., Chachaty C., Wolf C., Daveloose D., Bereziat G., Biochem. J., **264**, 633 (1989).

21- Nordio P.-L in "Spin labelling, Theory and Applications" Academic Press, N.Y. (1976).

22- Chachaty C., Can. J. Chem.,**62**, 1246 (1984).

23- Chachaty C., J. Chim. Phys., **82**, 621 (1985).

24- Olivier J.L., Chachaty C., Quinn P.J., Wolf C. J. Lipid Mediators (in press).

An Electron Spin Resonance Study of the Interactions Between Proteins and Lipids

S. AYNIE, M. LE MESTE

Laboratoire de Biologie Physico - chimique
Ecole Nationale Supérieure de Biologie Appliquée à la Nutrition et à l'Alimentation.
Campus Universitaire Montmuzard
21000 DIJON (FRANCE)

Despite the large using of milk proteins by the food industry, the mechanism of their action for emulsion stabilisation remains partially understood. Among the factors which are presumed to play a role in the mechanism, we selected to study both of them : 1) the affinity of milk proteins for lipids : neither its real importance, nor the nature of the lipids - proteins interactions are known. 2) the flexibility of protein which seems to have an essential function in the etablishment of the interactions with small solutes or polymers.

The interactions between proteins and lipids can be traduced by a change in the mobility of lipids molecules dispersed in solution or included in oil droplet, or of a labelled side chain of amino residues interacting after emulsification. Thus we based our approche on mobility mesurements using the electron spin resonance method (ESR). This technic is widely used to derive structural and dynamic informations about biological systems. These systems generally do not contain paramagnetic species required for this method. Also stable nitroxide radicals were added in solution or in emulsion :

- homologs of stearic acid (SA) shown in Figure 1: the 5- Doxyl- stearic acid (5SA) with the nitroxide moiety close to the polar head of the fatty acid, and the 16- Doxyl- stearic acid (16SA) with the nitroxyde close to the CH_3 end and reflecting the behaviour of the apolar chain.

B. Catoire (Editor)
Electron Spin Resonance (ESR) Applications in Organic and Bioorganic Materials

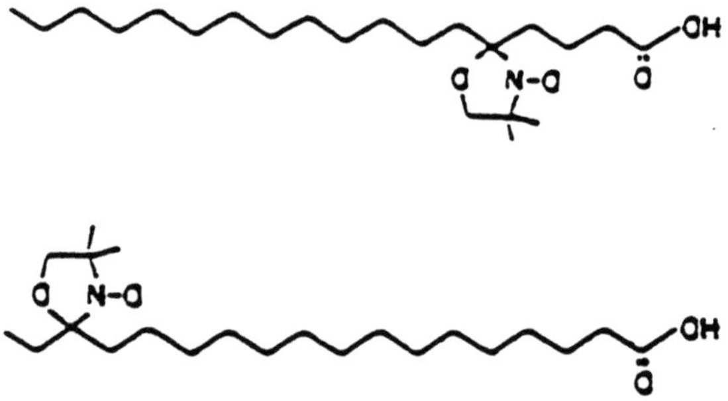

Figure 1

Nitroxides of stearic acid

- isothiocyanate nitroxide (Figure 2) covalently bound on the amino groups allowed to reveal either interactions with the lipid phase or modifications of the protein conformation.

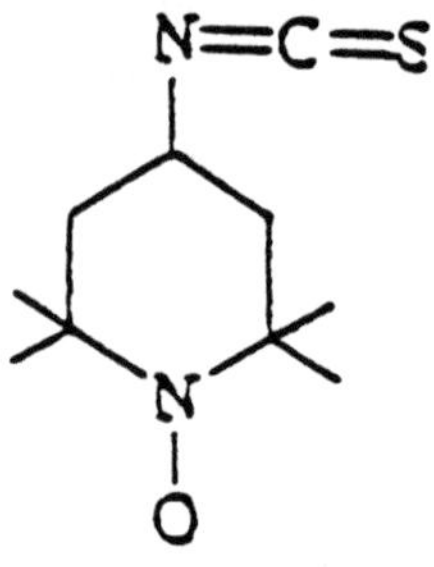

Figure 2

Isothiocyanato - nitroxide

First we compared the relative affinity of whey protein, whole casein or of their individual part for lipids; on the other hand, we assessed the contribution of the flexible side chains of proteins to these interactions.
Either 5SA or 16SA interactions with the main proteins were studied in parallel. The affinity of the protein for both probes being similar, changes on mobility of the nitroxide observed are depending only on their location onto the carbon chain.

Mobility of the polar end of stearic acid (5SA)

The spectra obtained with 5SA in buffer alone or in presence of whey protein are shown in Figure 3 a and b. The spectrum of 5SA in buffer reveals the presence of a rapid motion population and a well dispersion of the probe.

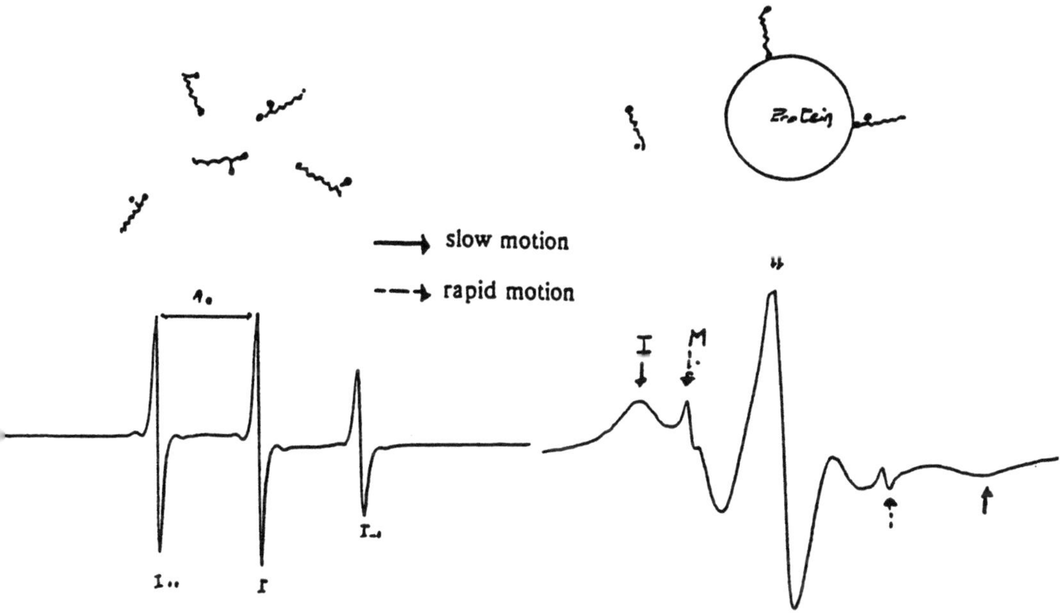

Figure 3

Spectra of 5- Doxyl- stearic acid in solution alone (a), upon interaction with whey protein (b).

When whey proteins were added (Figure 3b), a slow motion population I ($D_{rot} < 10^{-6} s^{-1}$) of nitroxide radical appeared. It corresponded to the powder spectrum. The comparison with simulated spectra allows to estimate that more than 95% of the radicals were assumed to interact with proteins.

Results given in Table 1 concerning R (=I/M), D_{rot} (rotational diffusion coefficient) and Ao values of the main milk proteins permitted to range the protein in order of R, i.e. of increasing affinity for the fatty acid in solution:

$$\alpha - lactalbumin < caseins < \beta - lactoglobulin = whey\_protein$$

Table 1

Comparison between milk proteins with 5SA in solution

5SA solution

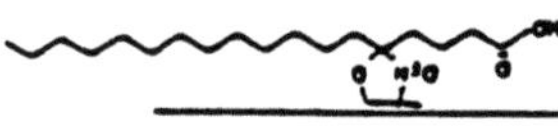

	$R = I/M$	D_{rot} (M) s^{-1}	A_0 (M) G
Buffer	0	1.10^9	16.4
α lactalbumin	0.3	$0.8\ 10^9$	15.7
β lactoglobulin	1.0	$0.5\ 10^9$	15.7
Whey proteins	1.0	$0.5\ 10^9$	15.7
α_s casein	0.4	$0.8\ 10^9$	15.7
β casein	0.4	$0.7\ 10^9$	15.8
Whole caseinate	0.4	$0.7\ 10^9$	15.8

FOR 16 SA $R \approx 0$

pH 7, buffer 10 mM (borate, citrate, phosphate, HCl)
protein concentration : 2.5 % (w/v)
nitroxide fatty acid : 0.25 mg/ml   molecular ratio Fatty acid / Protein = 0.5

D_{rot} precision ≈ ±5 %

$R \approx 1$ corresponds approximately to an immobilization of 98 % of the fatty acids (one fatty acid for 2 protein molecules)
$R \approx 0.4$ corresponds approximately to an immobilization of 75 % of the fatty acids

The high affinity of β – *lactoglobulin* for the fatty acid was confirmed at pH 5, pH at which this protein protected the fatty acid against aggregates whereas α – *lactalbumin* did not (Closs 1990). The good binding properties of β – *lactoglobulin* with apolar substances is well known (Spector et al. 1969; Papiz et al. 1986).

Whey proteins are globular and rigid whereas caseins are mainly random coils and flexible : thus it seems that the overall comformation of the protein and their overall flexibility are not determinant factors for their ability to interact with fatty acids. The particular rigidity of α – *lactalbumin* strongly reduced the accessibility of all its amino residues; this could explain its relatively low affinity for fatty acids. Even if the total hydrophobicity of these proteins is quite similar, the access to hydrophobic residues is higher for caseins. Thus the affinity for fatty acids could not be explained by this characteristic.

Mobility of the apolar chain (16SA)

In buffer, 16SA mobility was of the same order as 5SA mobility. In presence of the milk proteins, the mobility of the nitroxide close to the CH_3 end was not significantly modified. Moreover kinetic experiments conducted during 4 hours showed no differences between R at t=0 and R at t=4 hours (Aynié 1990).

It appears from these results that the interactions in solution between the milk proteins and fatty acids occured preferentially with the polar head of the fatty acid.

Behaviour of the fatty acid included into the oil droplet (Table 2)

In order to study the interactions between milk proteins and triglycerides in food emulsion, probes were also dispersed in the oil phase.

Upon the incorporation, the reduction of the nitroxide of the fatty acid mobility by a factor ten and Ao values were the confirmation of their presence at the surface of the fat droplet. The decomposition of the I-1 line indicates a partition of fatty acids betwen the aqueous and lipidic phases (Figure 4a ---).

Like in solution, whey protein added in emulsion strongly immobilzes 5SA; likewise 16SA was less sensitive (Table 2).

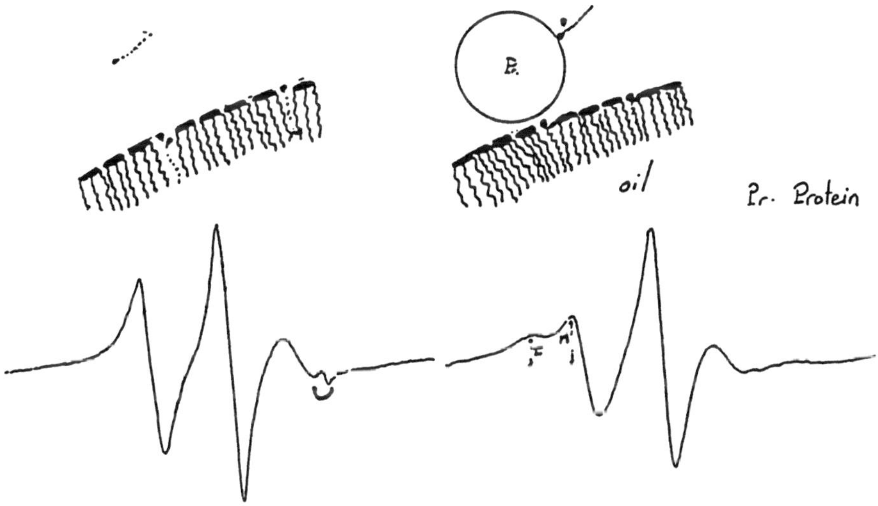

Figure 4

Spectra of 5- Doxyl- stearic acid in emulsion alone (a), upon interaction with whey protein (b).

Table 2

Values for different ESR parameters of 5SA and 16SA in solution and in emulsion with whey protein.

5SA Nitroxide (1 mg nitroxide/100 mg protein)

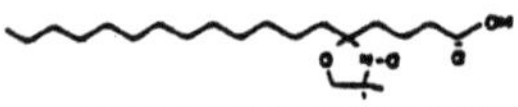	I/M	D_{rot} (M) s^{-1}	$A_{\parallel}$ (M) G
Buffer	0	1 10^9	16.4
Whey proteins	1.0	0.5 10^9	15.7
Emulsion (buffer)	0	1 10^8	14.5
Emulsion + whey prot.	0.7	7 10^7	14.5

16SA Nitroxide (1 mg nitroxide/100 mg protein)

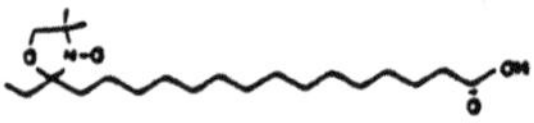	I/M	D_{rot} (M) s^{-1}	$A_{\parallel}$ (M) G
Buffer	0	1.6 10^9	15.7
Whey proteins	≃0	0.9 10^9	15.5
Emulsion (buffer)	0	1.6 10^8	14.5
Emulsion + whey prot.	0	2.3 10^8	14.5

Thus these results did not reflect a penetration of the whey protein inside the lipid monolayer : whey protein seems to bind to lipids of the oil as extrinsic proteins do to membrane phospholipids.

Motion of the protein side chains

This experiment was performed with whey protein and with its individual components.

Either with whey protein or with *β – lactoglobulin* in solution (Figure 5 and Table 3), two different populations can be distinguished : the M population can ascribed to the surface amino residues; the I one can be allocated to constrained motion radicals trapped within organized structures of the protein or within aggregates.

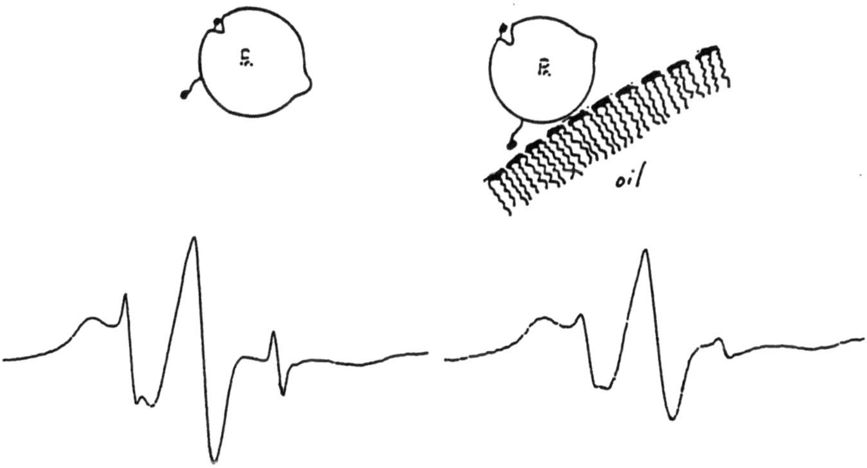

Figure 5

Spectra of spin labelled side chains of β-lactoglobulin in solution (a), and after emulsification (b).

On the opposite only a rapid motion population was observed with $\alpha -$ *lactalbumin*, probably because of a complete inaccessibility of buried amino groups which, in this way, can not be labelled.
The accessibility and mobility of lysyl residues of β-lactoglobulin were studied by Brown et al. (1983). It appeared three population of lysines with different mobility, corresponding to different locations upon $\beta -$ *lactoglobulin* (at the outer surface, near disulfide bonds or along the amphipatix helix). The spin-labelling results reflect this distribution of lysyl residues.

In emulsion, the rotational diffusivity of the more mobile chains was reduced (Table 3). This effect was more pronounced with $\alpha -$ *lactalbumin* . A decreasing of Ao values indicated the presence of both neighbouring proteins and interfacial lipids. Few spin labelled residues with slow motion were significantly affected by the interactions. Thus and similarly to Cornell and Patterson (1988), no modification of the conformation of the whey protein was observed when interactions occured. Moreover, recent studies on the flexible polar side chains of milk proteins (Le Meste et al., 1990) have stressed their possible involvement in the interactions milk protein - interfacial lipids.

Table 3

Mobility of the spin labelled amino residues of the proteins

In solution

	I/M	D_{rot} (M)	A_0 (M)
Whey proteins	0.3	1.1 10^9	16.9
α lactalbumin	≈ 0	2.6 10^9	17.1
β lactoglobulin	0.2	0.8 10^9	16.8

In the emulsion

	I/M	D_{rot} (M)	A_0 (M)
Whey proteins	0.3	0.2 10^9	15.6
α lactalbumin	0	0.3 10^9	15.4
β lactoglobulin	0.3	0.2 10^9	15.5

Opening of $\beta-$*lactoglobulin* and $\alpha-$*lactalbumin* thanks to chemical or biochemical modifications involved a significant improvement of the fatty acids - proteins interactions : higher accessibility of the proteins amino residues may be assumed to be responsable of this increasing of the interactions. So caseins glycosylation induced a better swelling of proteins, a more dissociated state (Le Meste 1990, Colas et al. 1988), but also a strong lowering of the affinity for the lipids (Courthaudon 1990), excepted for α_{s1} $-$*casein* which had less changed amino groups after this treatment.

All these experiments are in agreement with the assumption that accessible amino residues of milk proteins play an essential role in the interactions between these proteins and lipids.

ESR studies permitted to conclude that milk proteins - lipids interactions occured preferentially through their polar groups, making the more polar and accessible amino residues of the proteins, i.e. lysine involve. Any change in the conformation of the milk protein appears after emulsification. No close correlation between emulsifying properties of the milk proteins and their ability to interact with lipids in emulsion was found.

REFERENCES

Aynié S. (1990). DEA Sciences des Aliments, ENS. BANA, Dijon.

Brown E.M., Pfeffer P.E., Kumonsky T.F., Greenberg R. (1988). Biochemi stry, 27: 5601

Closs B. (1990). Thèse de Doctorat de l'Université de Bourgogne, ENS. BANA, Dijon.

Colas B., Gobin C., Lorient D. (1988). J. Dairy, 55: 539.

Cornell D.G., Patterson D.L. (1988). J. Agri. Food Chem., 37: 1455

Courthaudon J.L. (1990). Thèse de Doctorat de l'Université de Bourgogne, ENS. BANA, Dijon

Le Meste M., Colas B., Simatos D., Closs B., Courthaudon J.L., Lorient D. (1990) J. of Food Sci., In press.

Le Meste M., Closs B., Courthaudon J.L., Colas B. (1990). Macromolecular Interactions and Food Colloid Stability. In press.

Papiz M.E., Kraulis P.J., Eliopoulos E.E., North A.L.T., Findlay J.B.C., Sivraprasadarao R., Jones T.A., Newlomer M.E., Sawyer L. (1986). Nature, 324: 383.

Spector A.A, Fletcher J.E. (1969). Lipids, 5: 403

A Spin Probe ESR Study of Sugar Water Mixtures in the Liquid and Glassy State

M.J.G.W. ROOZEN, P. WALSTRA, T. VAN VLIET

Laboratory of Dairing and Food Physics, Department of Food Science

M.A. HEMMINGA

Department of Molecular Physics,
Wageningen Agricultural University, P.O. Box 8128 - 6700 ET Wageningen, the Netherlands

Abstract

Conventional and saturation transfer ESR spectroscopy are used to study the rotational behaviour of two different nitroxide spin probes: 4-hydroxy, 2,2,6,6-tetra-methylpiperidino-oxyl (TEMPOL) and 3-maleimido-2,2,5,5,tetra-methyl-1pyrrolodinyloxyl in glycerol water, sucrose water and dextrin water mixtures as a function of temperature. In the liquid state the results are discussed in terms of slip and stickiness of the spin probes, which depend on the strength of the hydrogen bonds between the spin probe and the solvent. Except for the malemeimido spin probe in anhydrous glycerol, which can form an extended solvation shell, the slip increases as the water content of the mixture decreases. This is explained by a decrease of hydrogen bonds between the spin probe and the solvent. In sucrose water mixtures above 70% by weight sucrose and in sucrose water mixtures in the glassy state the spin probes are presumably present in cavities. At the temperature at which the transition from glassy state to solution takes place an extreme increase in rotational mobility of the spin probes can be observed. The temperature at which this increase is observed agrees reasonably well with the glass transition temperature as determined by differential scanning calorimetry experiments. The apparent activation energy for rotation in glassy maltodextrin increases as the molecular weight becomes larger, presumably due to a decrease in the free volume of the sample. No influence of the water content of the glassy samples at temperatures above the freezing point of water on the apparent activation energy is observed.

B. Catoire (Editor)
Electron Spin Resonance (ESR) Applications in Organic and Bioorganic Materials

Introduction

During the last years there has been an increasing interest in systems in the glassy state. [1-8] A glass is an amorphous solid, which exhibits a glass transition.[9] In amorphous materials the molecules are forming a non-periodic and non-symmetric network.[10] The glass transition is the phenomenon in which a solid amorphous phase exhibits a discontinuous change in the specific heat on changing the temperature. An essential prerequisite for glass formation from a solution is that the cooling rate must be sufficiently fast to preclude nucleation and crystal growth.[9] At the glass transition temperature, there is not only a sudden change in thermal and mechanical properties of the system, but also an extreme decrease in the rates of molecular translation diffusion.[11] Due to very limited molecular motion a food product in the glassy state is presumed not to decrease in quality during storage.

The temperature (T_g) at which carbohydrate water mixtures have their glass transition depends strongly on the water content of those mixtures. If the glass transition takes place at temperatures below the freezing point of the solution, a part of the water may crystallize. The viscosity of the remaining concentrated solution will become so high that eventually freezing of the remaining water is inhibited[11]. The physical state of such a mixture consists of ice crystals embedded in a solid supersaturated matrix. The temperature at which water stops freezing, and were also a discontinues change of the specific heat takes place is the special glass transition temperature (T_g'). The amount of water remaining unfrozen is often assumed to be independent of the starting concentration[3]. At rewarming of such a mixture, the Stokes viscosity of the amorphous solution starts to decrease strongly above T_g' due to the melting of ice and the increase of the free volume[6].

To obtain a better understanding of the molecular dynamics in the glassy state, spin probe electron spin resonance spectroscopy (ESR) has been shown to be a good method to study molecular motion in bio-organic glasses[12, 13]. ESR spectroscopy gives information about the rotational mobility of spin probes that are dissolved in the systems to be examined.

In literature several applications of spin probe ESR in glassy systems have been described. By using both conventional and saturation transfer ESR, Ito[14] demonstrated for several low molecular organic systems that the change of the matrix due to the glass transition is reflected in a decrease of the rotational correlation time of nitroxide spin probes. Hamada et al.[15] have shown that the rotational correlation time of different nitroxide spin probes in nylon 6 films decreased with rising water content at a given temperature. Simatos[16] discussed the rotational mobility of spin probes in water dextran mixtures. For water activities below 0.75 no spin probe mobility could be detected on a time scale of 10^{-7} s or less. Le Meste and Duckworth[12] studied the mobility of spin probes in concentrated caseinate solutions at room temperature. They observed an increase of rotational mobility above a water content of 0.25 g water/g dry protein, which was ascribed to a glass-rubber transition.

The present paper describes the application of conventional and saturation transfer ESR [17, 18] to study the rotational mobility of spin probes in sucrose water mixtures and various maltodextrin water mixtures above and below the glass transition temperature. By using conventional ESR as well as saturation transfer ESR rotational mobilities between 10^{-3} and 10^{-11} s could be determined. As described by Jolicoeur[19] and Le Meste and Voilley[20] glycerol water mixtures were used as a reference systems for this study.

Theory

To describe the isotropic rotational motion for small molecules in liquids, a modified Stokes-Einstein relationship can be successfully employed:[13, 21-24]

$$\tau_c = \{\eta V/k_b T\}\, k + \tau_0, \qquad (1)$$

where τ_c is the rotational correlation time, η is the solvent viscosity, k_b is Boltzmann's constant, V the volume of the rotating molecule, T the absolute temperature, τ_0 is the zero viscosity rotational correlation time and k a dimensionless interaction parameter. The parameter k is a measure of the coupling of the rotational motions of the spin probe to the shear modes of the fluid. It has been found in many cases that $0 \leq k \leq 1$ and that k is independent of temperature and viscosity. The parameter k depends, however, both on specific solvent probe interactions and on the geometry of the molecules considered, since non-spherical molecules must "displace" solvent as they rotate. This can be expressed as follows:[25]

$$k = S\,(1 - \theta) + \theta, \qquad (2)$$

in which S is the stickiness factor that depends on the coupling between the rotating molecule and the liquid. If S is zero, there is no interaction (complete slip conditions); for S equals one the boundary layer conditions are completely sticking. The parameter θ is the ratio of the effectiveness of the torques under slip to those under stick conditions, and it depends on the geometry of the rotating molecule. Values derived from hydrodynamic theory have been tabulated for prolate and oblate spheroids as a function of the ratio of the short and long molecular axes.[26] For the spin probes used in this study these values were determined by means of Stuart Briebleb atomic models.[13]

For non-spherical molecules, S can be negative, implying that the molecules displace less solvent than is expected from their geometry. This is the case, for example, if a cavity is formed in which the spin probe can carry out "free rotation". k is then equal to zero, so that S equals $-\theta/(1 - \theta)$ (see Equation 2).

Materials and methods

Sample preparation

Glycerol solutions were prepared by mixing 87% glycerol (Merck) and spin probe (Aldrich) solution. Concentrated glycerol solutions were prepared by drying the mixture at 102 °C. Anhydrous glycerol was obtained by drying during 60 hours. Sucrose solutions were made by mixing sucrose (Merck) and spin probe solution during heating. The water content was determined from the refractive index as tabulated by Weast.[27] Nitroxide radicals I (4-hydroxy, 2,2,6,6-tetra-methyl-piperidino-oxyl (TEMPOL)) and II (3-maleimido-2,2,5,5,tetra-methyl-1-pyrrolodinyloxyl) were used as spin probes. The final spin probe concentration in the samples was 0.2-0.5 mg/ml.

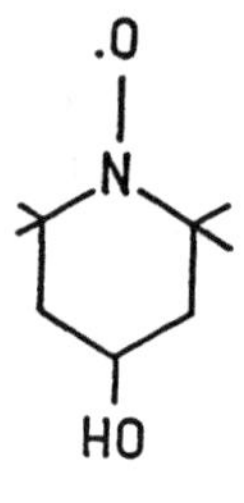

Schema I

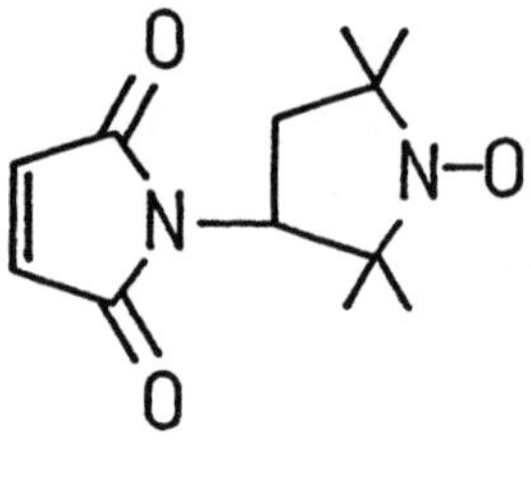

Schema II

Maltodextrines solutions were prepared by mixing dry maltodextrin (tri-, penta-, and heptasaccharides; Boehringer) with an aqueous solution of spin probe I (TEMPOL). To obtain concentrated maltodextrin water mixtures, the solutions were dried for two weeks above concentrated salt solutions as described by Murray.[28] The samples were ground in a mortar and dried again for a week. The dried samples were rapidly brought into capillaries (diameter: 1 mm), which were sealed to prevent water uptake. The sample length was 40 mm. The water content of the samples was determined by drying for 16 h at 102 °C.

Spectroscopy

ESR spectra were recorded on a Bruker ESR spectrometer 200D with nitrogen flow temperature control. For conventional ESR the microwave power was 1-5 mW. The scan range, scan rate, time constant and modulation amplitude were adjusted so that distortion of the spectra was avoided. The rotational correlation time (τ_C) of weakly immobilized spin probes ($\tau_C < 10$ ns) was estimated from the relation:[17, 29]

$$\tau_C = 6.5\times10^{-6} \times B_0 \left(\sqrt{h_H/h_C} - 1\right),$$

where h_H and h_C are the heights of the high field and central lines in the ESR spectra, respectively. B_0 is the linewidth of the central line in tesla (T). The rotational motion of the spin probes is assumed to be isotropic.

In the slow motional region of the spin probes (τ_C between 10 and 200 ns) the rotational correlation time was obtained by using the method of Goldman:[17, 30]

$$\tau_C = a(1 - A'_Z/A_Z)^b,$$

in which A'_Z is the separation of the outer hyperfine extremes in the ESR spectra, and A_Z is the rigid limiting value for the same quantity. Both a and b depend on the nature of the motion of the spin probe and of the intrinsic linewidth of the spectra. The Brownian diffusion model has been taken, resulting in: $a = 1.09\times10^{-9}$ s and $b = -1.05$.[13, 31, 32]

Saturation transfer ESR is applied in the very slow motional region (10^{-7} s < $\tau_c < 10^{-3}$ s). Spectra were recorded under saturation conditions: microwave power 200 mW, and a modulation amplitude of 1 mT. The ESR signal was recorded in quadrature[18] with respect to the modulation signal. The rotational correlation time τ_c was estimated by comparing the recorded spectra with standard spectra[17, 18, 33], recorded in anhydrous glycerol under similar conditions as for the mixtures. Values of τ_c between 10^{-7} and 10^{-3} s are accurate within a factor of 2.[33]

For the samples that were studied at temperatures below the freezing point of water, the system was cooled rapidly to -70 °C and rewarmed to the temperature at which the spectrum was recorded. This ensures maximally frozen samples.[34]

Results and discussion

Glycerol water mixtures

To analyze the interaction between the spin probe and the solvent in glycerol water and sucrose water mixtures at temperatures above the freezing point, graphs of the values for τ_c versus η/T were plotted (see Equation 1; data not shown). Values for η for sucrose water mixtures are obtained from Bates[35] and for glycerol water mixtures from Segur.[36] To obtain values for k, from the slope of these graphs, the volume of the spin probe is considered to be independent of the composition of the solution. The Van der Waals incremental volumes as tabulated by Bondi[37] are used.[13] In the temperature range considered here, k was found to be independent of temperature. The intercepts of these graphs (τ_0) are between 10^{-10} and 10^{-12} s, in agreement with the results of other studies.[22]

Parameter S describes specific solvent probe interactions, which are independent of the shape of the spin probe. To obtain values for S from Equation 2, values for θ are determined from the molecular shape of the spin probes.[13] To study variations of S with solvent composition, the value for θ is considered to be independent of temperature and solvent effects.

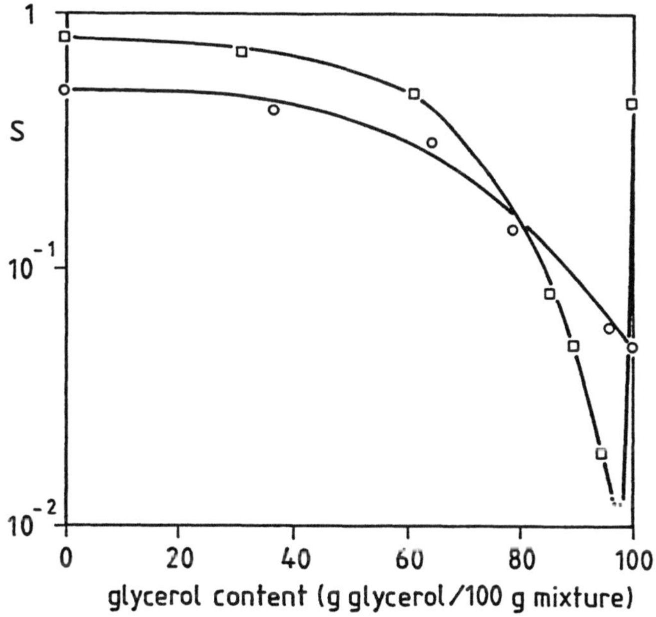

Figure 1. Parameter S as a function of the glycerol content for glycerol water mixtures for spin probe I (O) and spin probe II (□).[13] (Reproduced with permission)

Figure 1 shows parameter S as a function of the glycerol content in different glycerol water mixtures. For both spin probes a decrease of S with increasing glycerol content is observed up to a glycerol content of 95%. This result agrees with results of Le Meste and Voilley.[20] The decrease of S implies that both spin probes will interact stronger with water than with glycerol molecules. Therefore at increasing glycerol contents the slip will increase. This is in agreement with the statements of Douzou.[38] Douzou stated that glycerol is sufficiently more basic in character than water to behave as a hydrogen acceptor, rather than as an amphoteric compound. Spin probe I is a secondary alcohol, so that it slightly prefers to accept a hydrogen bond rather than to offer it.[38] Spin probe II is a di-ketone and is a stronger base than spin probe I. The decrease of S with glycerol content in aqueous glycerol mixtures is therefore in agreement with the acid-base aspects of the formation of a hydrogen bond.

Above 95% glycerol, an extreme increase in S of spin probe II is observed. This is not the case for spin probe I, which shows a continuous decrease up to anhydrous (about 100%) glycerol. Spin probe I hardly interacts with anhydrous glycerol. Spin probe II, however, interacts strongly with the same solvent. This spin probe will, due to its relatively strong basic character, form a solvation shell with the amphoteric glycerol molecules. If a small amount of water is added to anhydrous glycerol, the glycerol molecules in the solvation shell of spin probe II will be replaced by the smaller water molecules. This leads to a decrease of its hydrodynamic volume and explains the minimum of parameter S at low water contents.

The observed differences between the values for parameter S of the two spin probes at low glycerol contents are relatively small. Since S is independent of the molecular shape of the spin probes, this can be due to a different dynamic behaviour of the spin probes (i.e. anisotropic and internal motion) as well as to uncertainties in the values selected for V and 0.

Sucrose water mixtures

In Figure 2 the parameter k is given as a function of the sucrose content in sucrose water mixtures. A logarithmic plot of S (as in Figure 1) cannot be given here, since S becomes negative. The value of k at which this occurs is indicated with arrow 1 in Figure 2. At concentrations above 40% sucrose a decrease of k is observed, this point is indicated with arrow 2.

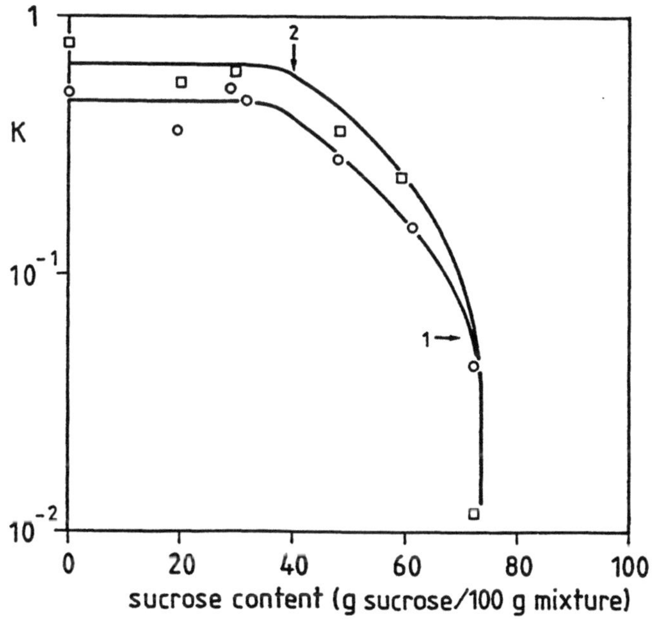

Figure 2. Parameter k as a function of the sucrose content in sucrose water mixtures for spin probe I (O) and spin probe II (□). Arrow 1 indicates the value for k where S becomes negative. Arrow 2 indicates the sucrose content where k starts to decrease.[13] (Reproduced with permission)

In Figure 2, it can be seen that the value of parameter k in sucrose water mixtures between 0 and 40 % by weight (= 3.4 mol H_2O/mol) is constant and close to 1. This is explained by the fact that the spin probe molecules can form many hydrogen bonds with the liquid. Due to the relative basic character of the spin probes, they will form hydrogen bonds with water rather than with sucrose molecules. It is known from other studies that the hydrogen bonds between the sucrose and water molecules in an aqueous sucrose solution are stronger or more extensive than the bonds between water molecules themselves.[39, 40]

At concentrations above 40% sucrose a decrease of parameter k is observed (arrow 2 in Figure 2). This is explained by the fact that fewer hydrogen bonds are formed between the spin probes and the solvent. This is in agreement with the conclusions of Flink,[10] that sucrose water mixtures at concentrations over 30-40% sucrose change from a solution of hydrated sucrose molecules to a sucrose-water phase, in which all the water molecules directly or indirectly (i.e. as a second layer) are involved in hydrogen bonds with sucrose.

Arrow 1 in Figure 2 shows that parameter k equals θ (and thus S becomes negative) at about 70% by weight sucrose (= 11 mol H_2O/ mol). A negative value of S is indicative for the the formation of cavities. Cavities can be formed if the spin probes are not part of the lattice. It may be possible, however, that the cavities are not a property of the lattice, but are induced by the spin probes itself.

The glass transition temperature

The rotational correlation time for both spin probes after rewarming of a rapidly cooled 20% sucrose water mixture is shown in Figure 3. Between -70 and -31 °C log τ_c decreases almost linearly with temperature. At temperatures between -33 and -28 °C (as indicated with arrows in Figure 3) τ_c starts to decrease much stronger.

According to experimental observations in our laboratory (data not shown) at temperatures where a part of the water is frozen, the spin probes are not present in the ice lattice, but in the concentrated amorphous solution. This is in agreement with results of other studies.[41] The strong increase of mobility of the spin probes that starts between -33 and -28 °C is explained by the glass transition of the mixture, which is at -32 °C.[42, 43]

Due to the glass transition the Stokes viscosity η of the concentrated amorphous solution decreases. The reduction of η for a 20 % sucrose water is from about 10^{12} Pa•s at the glass transition temperature to about 10^3 Pa•s at temperatures 20 °C above the glass transition.[6] The relative decrease of τ_c, as shown in Figure 3 is, however, much smaller. It should be noted that a 20% sucrose water mixture at temperatures below the glass transition consist of ice and a glassy sucrose water mixture. When the ice starts to melt due to the glass transition parameter k will increase strongly as can be seen from Figure 2. This effect will reduce the decrease of τ_c due to the glass transition.

At -50 °C the rotational correlation time is about 10^{-4} s (Figure 3), so that k is about 10^{-9} (given a viscosity of 10^{12} Pa•s[11]). This very low value of k implies that the spin probes are present in cavities in the lattice of the amorphous solution, not being part of the lattice. Spin probe II is somewhat more mobile in the glassy sucrose system than spin probe I. Possibly, the distortion of the lattice due to the presence of spin probe I is smaller than that due to the presence of the less spherical spin probe II.

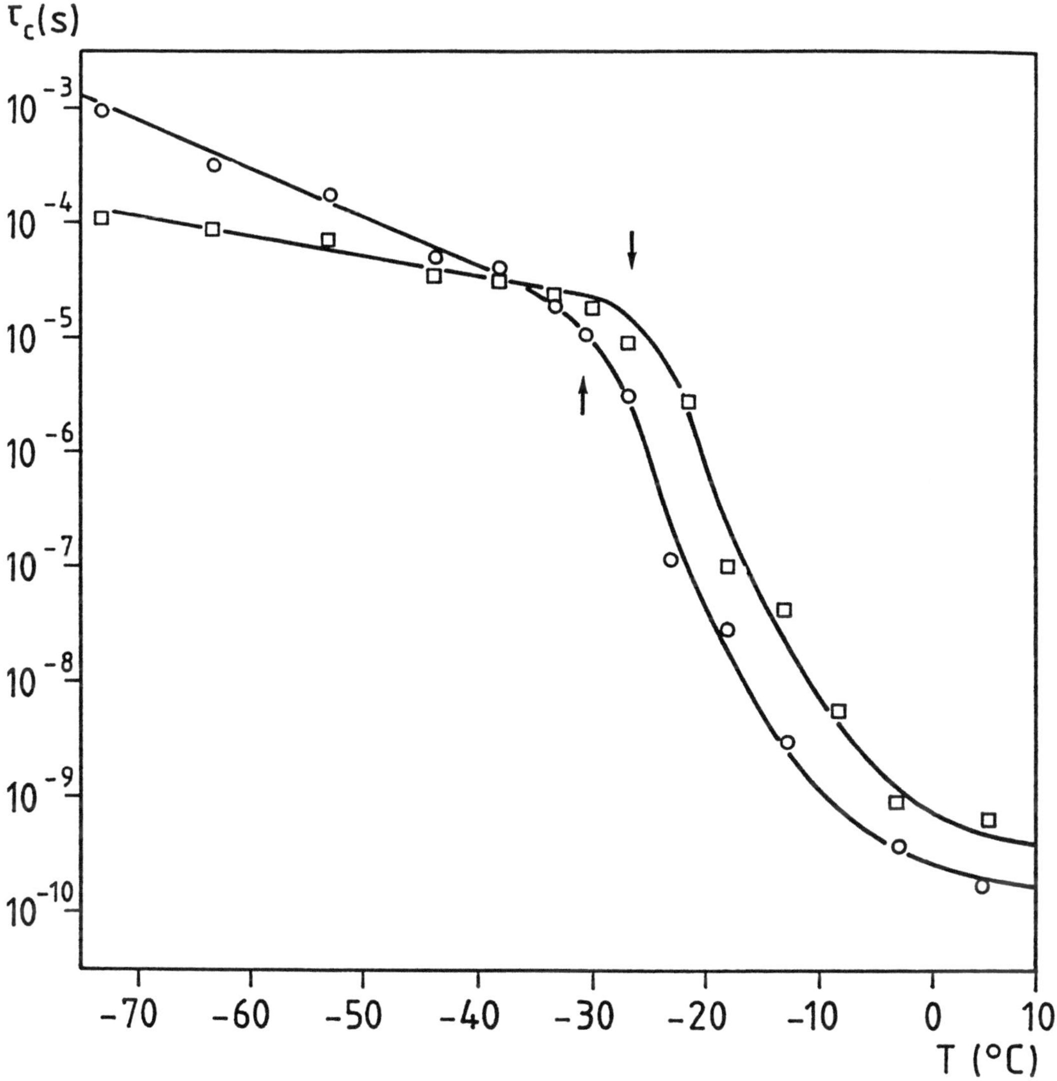

Figure 3. Rotational correlation time τ_c as a function of temperature in a 20% sucrose water mixture for spin probe I (O) and spin probe II (□). The temperature at which τ_c starts to decrease sharply is indicated with an arrow.[13] (Reproduced with permission)

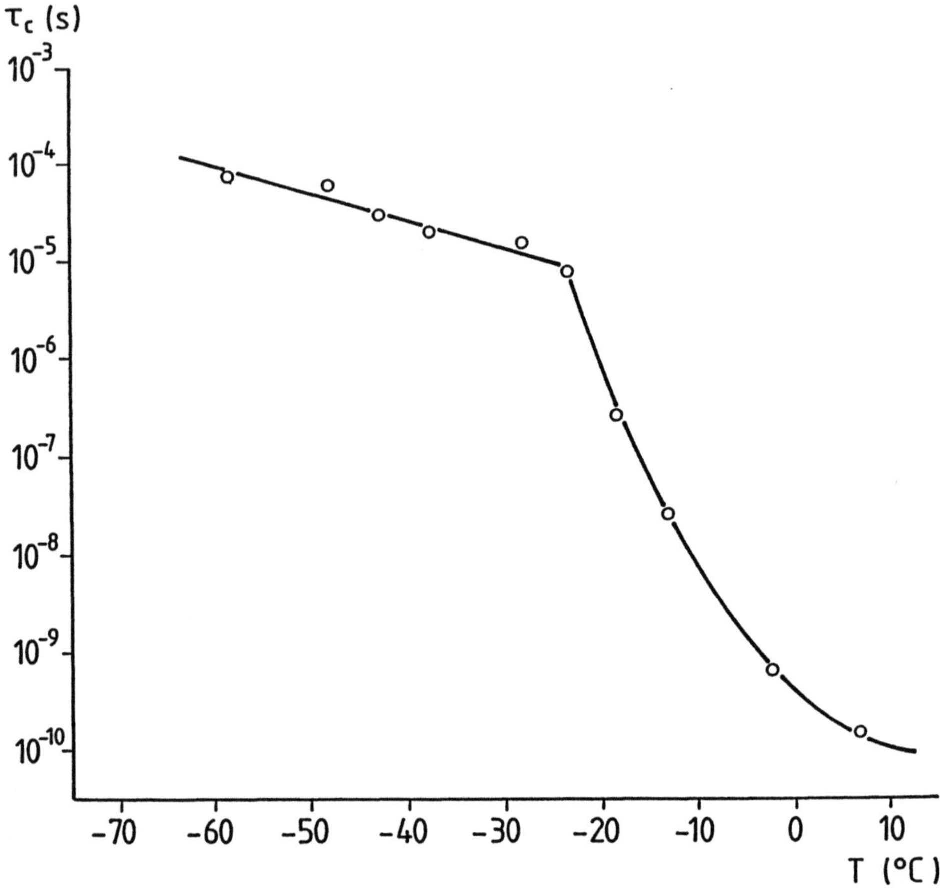

Figure 4. Rotational correlation time τ_c of TEMPOL spin probe I as a function of temperature in a 20% maltotriose water mixture.[44] (Reproduced with permission)

Figure 4 shows a plot of τ_c of spin probe I versus temperature at rewarming of a rapidly cooled 20% maltotriose water mixture. In this graph a sudden change in the temperature dependence of the rotational mobility can clearly be observed at a temperature of -23 °C. This temperature is very close to the special glass transition temperature T_g' as tabulated by Levine and Slade[43] (T_g'= -23.5 °C). Similar effects are obtained for other maltodextrin solutions.[44]

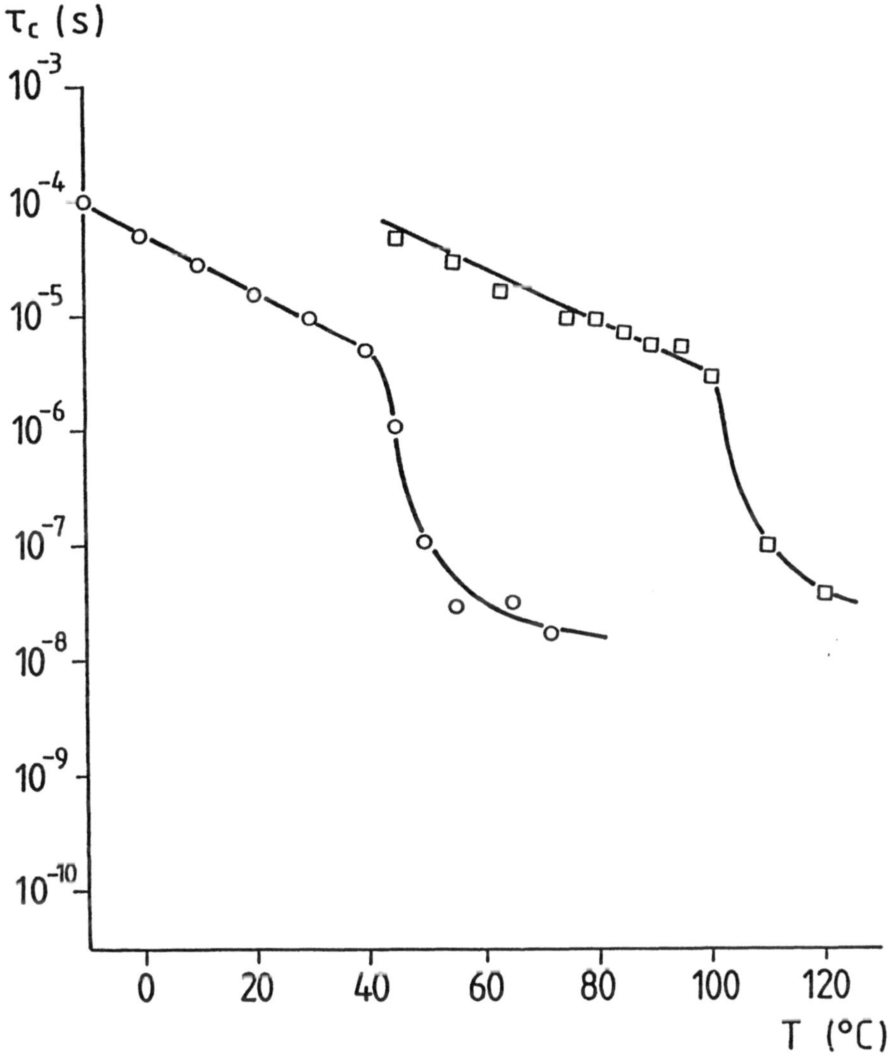

Figure 5. Rotational correlation time τ_c of TEMPOL spin probe I as a function of temperature in a maltoheptaose water mixture stored at an equilibrium relative humidity of 33% (O) and oven dried for 16 hours at 100 °C(□).[44]
(Reproduced with permission)

Figure 5 shows the rotational mobility of spin probe I in a concentrated maltoheptaose water mixture (equilibrium relative humidity: 33%). This graph is identical in shape to Figure 4; the glass transition (T_g) can be observed at 42 °C. Due to a higher viscosity the value of τ_c at temperatures above T_g in Figure 5 is higher than in the samples in which ice melts (Figure 4). In addition, Figure 5 shows results for an oven dried maltoheptaose sample (16 h at 102 °C). Due to the decrease of the water content, T_g is shifted upwards to about 100 °C.

Commercial maltodextrin

To compare our results with those on materials used in the food industry, such as enzymatically converted starch, which is a mixture of dextrin molecules of different molecular weight, a commercially available sample of maltodextrin was examined: Passelli SA 2, with an average molecular weight of 8000 Da. In Figure 6 the rotational mobility of spin probe I in this sample (ERH: 57%) is shown as a function of temperature, giving a sudden change in the temperature dependence of the mobility at 30 °C.

It is also shown in Figure 6 that adding of 8% (by weight) glucose to the polymer results in a decrease of the glass transition temperature by 5 °C. Several effects could contribute to this decrease: (1) the addition of glucose results in a small increase of the water to dextrin ratio; (2) glucose might be able to act as a plasticizer to dextrin, although on the other hand water molecules that interact with glucose molecules may not be able to plasticize the polymer chain.

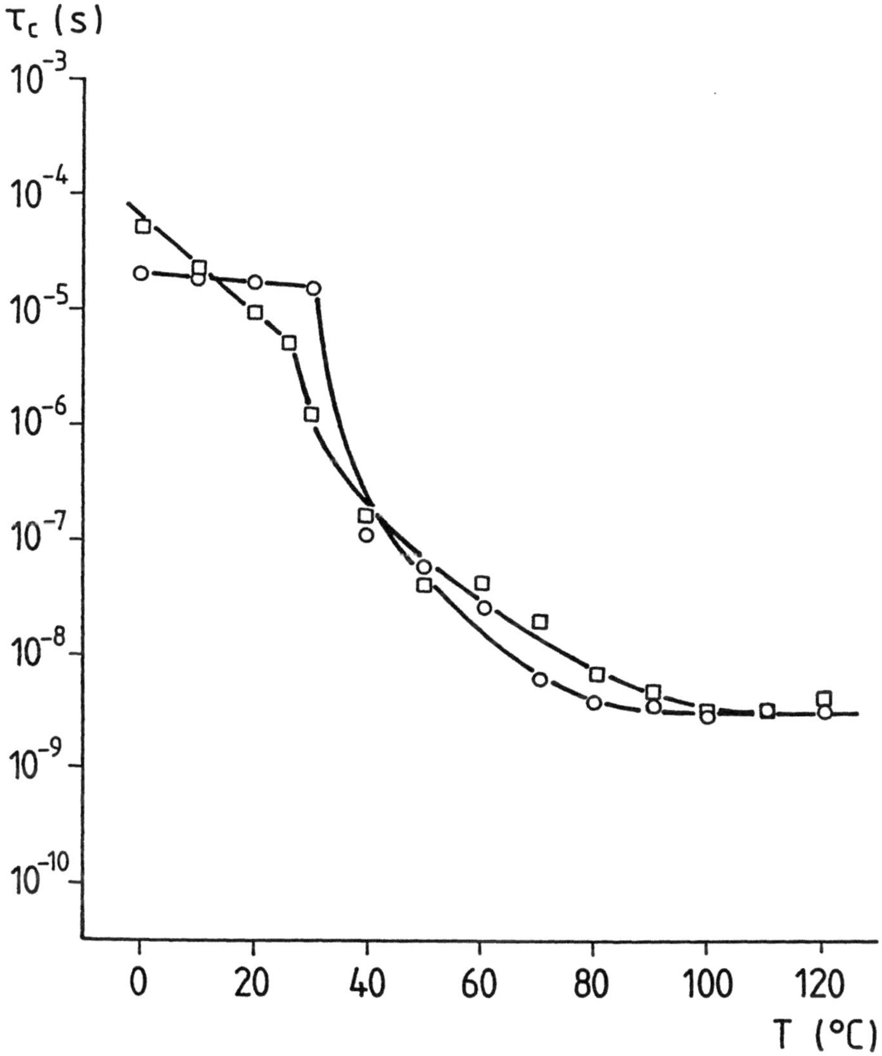

Figure 6. Rotational correlation time τ_c of TEMPOL spin probe I as a function of temperature in a heterodispers maltodextrin polymer (Passelli SA 2) mixture with 12.7% water (O) and with 12.7 % water and 8% glucose (□).[44] (Reproduced with permission)

Rotational mobility in glassy systems

The temperature dependency of the correlation time of spin probes is frequently described by an Arrhenius type equation:

$$\tau_c = \tau_a \exp(E_a/RT), \tag{3}$$

where τ_a is a temperature independent factor, R and T have their usual meaning; E_a is often interpreted as an activation energy for rotation[14, 15, 45, 46]. The values of E_a for the spin probe I in glassy maltodextrin water mixtures are tabulated in Tables 1 and 2.

Table 1

Apparent activation energy E_a for rotation of TEMPOL spin probe I in glassy 20% maltodextrin water ice mixtures

Maltodextrin	E_a (kJ/mol)
maltotriose	17
maltopentaose	14
maltoheptaose	26

Table 2

Apparent activation energy for rotation of TEMPOL spin probe I in concentrated glassy maltodextrin water mixtures

Maltodextrin	water content (%)[a]	E_a (kJ/mol)
maltotriose	2.5	35
maltotriose	10.0	49
maltotriose	11.5	35
maltotriose	19.5	41
maltopentaose	4.0	35
maltopentaose	12.5	26
maltopentaose	13.5	26
maltopentaose	18.5	43
maltoheptaose	3.5	21
maltoheptaose	11.5	29
maltoheptaose	14.5	34
maltoheptaose	18.5	34

[a]determined by drying at 100 °C overnight, corrected by +2.5%, +4% and +3.5% for respectively maltotriose, maltopentaose and maltoheptaose, respectively.[44]

As discussed before, spin probes in glassy sucrose water mixtures are presumably present in cavities in a lattice of the amorphous solution. The value for τ_c in glassy maltotriose (Figure 4) indicates that there are also cavities in glassy maltodextrines. However, the absolute values for τ_c and E_a clearly show that the spin probe I does not rotate freely in such cavities.

Two important factors have been proposed to determine the rotational mobility of spin probes in glassy systems: (1) The free volume of the "host" systems;[47] (2) Specific interaction between the spin probe and the chain molecule.[15] Hydrogen bonds often are the most important interaction.[13, 47] Both factors do not only influence the absolute value of τ_c, but also the change of τ_c as a function of temperature (as expressed by E_a). In Table 2, it can be seen that the values for E_a in maltotriose samples are probably somewhat higher than in the maltopentaose and maltoheptaose samples. This may be explained by an increase of free volume with increasing molecular weight.[48] The same effect is more pronounced in Figure 6: addition of glucose to the polymer presumably results in a decrease in the free volume of the system, which may explain the strong increase of E_a from 8 kJ/mol to 60 kJ/mol.

The value for E_a in glassy maltodextrin water mixtures (between 21 and 49 kJ/mol) is relatively high as compared to the results from studies of spin probes in glassy nylon water films.[46, 49] In the latter system, the spin probes appear to rotate in local defects with a low apparent activation energy of 2-8 kJ/mol and a rotational correlation time between 10^{-8} and 10^{-7} s. The relatively high value for both E_a and τ_c, as found in this study, will probably be due to the difference in number and strength of hydrogen bonds formed between the spin probe and its surrounding. Also the relatively small free volume of the maltodextrin, due to its small molecular weight as compared to nylon, may add to this effect.

Taking into account that the uncertainty in E_a in Table 2 is at least 10%, no significant influence of the water content of the samples on E_a can be observed. The values for E_a in frozen maltodextrin water mixtures as shown in Table 1 are smaller than in the concentrated samples (Table 2). This is explained by the possibility that the presence of ice crystals may cause an increase of the free volume or might reduce the interaction with the spin probe.

Conclusions

From our experiments it can be concluded that the changes in macroscopic observable properties of sugar water mixtures as determined by means of rotational diffusion measurements of ESR spin probe molecules, agree very well with differential scanning calorimetry measurements. ESR is shown to be a powerful technique to obtain information about hydrogen bonds, mobility and formation of cavities in glassy and liquid sugar water systems.

Acknowledgements

We are grateful to Unilever Research, Colworth House, UK, for financial support.

References

1. Franks, F. in *Properties of water in foods* (eds. Simatos, D. & Multon, J.) 497 (Nijhoff publishers, Dordrecht,, 1985).

2. Biliaderis, C.G., Page, C.M., Maurice, T.J. & Juliano, B.O. *J. Agr. Fd. Chem.* **34**, 16 (1986).

3. Levine, H. & Slade, L. in *Food Structure: its Creation and Evaluation* (eds. Blanshard, J.M.V. & Mitchell, J.R.) 149 (Butterworths, London, 1988).

4. Slade, L. & Levine, H. in *Food Structure: its Creation and Evaluation* (eds. Blanshard, J.M.V. & Mitchell, J.R.) 115 (Butterworths, London, 1988).

5. Simatos, D. & Karel, M. in *Food Preservation by Moisture Control* (eds. Seow, C.C.) 1 (Elsevier Applied Science, UK, 1988).

6. Simatos, D., Blond, G. & Le Meste, M. *Cryoletters* **10**, 77 (1989).

7. Blond, G. *Cryo-Letters* **10**, 299 (1989).

8. Orford, P.D., Parker, R., Ring, S.G. & Smith, A.C. *Int. J. Biol. Macromol.* **11**, 91 (1989).

9. Elliot, S.R. *Physics of amorphous materials* (Longman, London, 1983).

10. Flink, J.M. in *Physical Properties of Foods* (eds. Peleg, M. & Bageley, E.B.) 473 (AVI publishing Co. Inc., Westport, Ct, 1983).

11. Franks, F. *Biophysics and Biochemistry at low temperatures* 1-37 (Univ. Press, Cambridge, 1985).

12. Le Meste, M. & Duckworth, R.B. *Int. J.Food. Sc. Technol.* **23**, 457 (1988).

13. Roozen, M.J.G.W. & Hemminga, M.A. *J.Phys.Chem.* In press, (1990).

14. Ito, Y. *J. Chem. Phys.* **79**, 2650 (1983).

15. Hamada, K., Iijima, T. & McGregor, R. *Macromolecules* **19**, 1443 (1986).

16. Simatos, D., Le Meste, M., Petroff, D. & Halphen, B. in *Water Activity: Influences on Food Quality* (eds. Rockland, L.B. & Stewart, G.F.) 319 (Academic, Orlando, Fl, 1981).

17. Hemminga, M.A. *Chem. Phys. Lipids* **32**, 323 (1983).

18. Hemminga, M.A. & de Jager, P.A. in *Biological Magnetic Resonance. Spin Labeling - Theory and Applications* (eds. Berliner, L.J. & Reuben, J.) 131 (Plenum Press, New York, 1989).

19. Jolicoeur, C.J. & Friedman, H.L. *Ber. Bunsen-Ges. Phys. Chem.* **76**, 248 (1971).

20. Le Meste, M. & Voilley, A. *J. Phys. Chem.* **92**, 1612 (1988).

21. McClung, R.E.D. & J., K.D. *Chem. Phys.* **59**, 3380 (1968).

22. Kivelson, D., Kivelson, M.G. & Oppenheim, I. *J. Chem. Phys.* **52**, 1811 (1970).

23. Kowert, B. & Kivelson, D. *J. Chem. Phys.* **64**, 5206 (1976).

24. Dote, J.L., Kivelson, D. & Schwartz, R.N. *J. Phys. Chem.* **85**, 2169 (1981).

25. Hoel, D. & Kivelson, D. *J. Chem. Phys.* **62**, 1323 (1975).

26. Hu, C.M. & Zwanzig, R. *J. Chem. Phys.* **60**, 4345 (1974).

27. Waest, C. *Handbook of Chemistry and Physics* (Chemical Rubber Co, 1970).

28. Murray, D.G. & Luft, L.R. *Food Technol.* **3**, 32 (1973).

29. Knowles, P.F., Marsh, D. & Rattle, H.W.E. *Magnetic Resonance of Biomolecules* (John Willey & Sons, London, 1976).

30. Freed, J.H. in *Spin labeling, theory and applications* (eds. J., B.L.) 53 (Academic Press, New York, 1976).

31. Hyde, J.S. in *Methods in Enzymology* (eds. Hirs, C.H.W. & Timashef, S.N.) 480 (Academic Press, New York, 1978).

32. Le Meste, M. & Simatos, D. *Cryo-letters* **1**, 402 (1980).

33. Hyde, J.S. in *Molecular Motions in Polymers* (eds. Boyer, R.F. & Keinath, S.E.) 287 (Academic Press, New York, 1978).

34. Levine, H. & Slade, L. *J. Chem. Soc. Faraday. Trans. I* **84**, 2619 (1988).

35. Bates, F.J. *Polarimetry, Saccharimetry and the Sugars* (Natl. Bur. Standards, Washington D.C., 1942).

36. Segur, J.B. & Oberstar, H. *Ind. & Eng. Chem.* **43**, 2117 (1951).

37. Bondi, A. *J. Phys. Chem.* **68**, 441 (1964).

38. Douzou, P. *Cryobiochemistry: An introduction* (Academic Press, Orlando, Fl, 1977).

39. Taylor, J.B. & Rowlinson, J.S. *Trans. Faraday. Soc.* **51**, 1183 (1955).

40. Sugett, A. *J. Sol. Chem.* **5**, 33 (1976).

41. Yoshioka, H. *Chem. Letters* 1153 (1977).

42. McKenzie, A.P. *Phil. Trans. R. Soc. London.* **278**, 16 (1977).

43. Levine, H. & Slade, L. *Cryo-letters* **9**, 21 (1988).

44. Roozen, M.J.G.W., Hemminga, M.A. & Walstra, P. *Carbohydr. Res.* **In press**, (1990).

45. Hamada, K., Iijima, T. & McGregor, R. *J. Polymer Sci.: part B. Polym. Phys.* **25**, 1299 (1987).

46. Hamada, K., Iijima, T. & McGregor, R. *Polym. J.* **17**, 1245 (1985).

47. Sandreczki, T.C. & Brown, I.M. *Macromolecules* **21**, 504 (1988).

48. Billmeyer, F.W. *Text book of polymer Science* 1-228 (Wiley-Interscience, New York, 1971).

49. McGregor, R., Iijima, T., Sakai, T., Gilbert, R.D. & K. Hamada, J. *Membrane Soi.* **18**, 129 (1984).

Spin Labelling and Membrane Fluidity

J.VIRET and F.LETERRIER

Unité de Biophysique du C.R.S.S.A.

38702 LA TRONCHE CEDEX

--=oo00&00oo=--

I - NOTION OF MEMBRANE FLUIDITY

In 1972, Singer & Nicolson (1) have introduced the expression of "fluid mosaic" in order to describe the biological membranes. Nowadays, and in a more functional than descriptive aim, the term "membrane fluidity" is rather used to express the structural organization of these membranes we notice that :

One of the ruling stakes of biology is to explain the membrane functioning. It can be done by showing up direct correlations between their detailed biochemical composition, their more of less fluid structural state and their functionality. Our purpose does not consist in developping here neither the results of these researches nor in discussing their conceptual validity. It is enough to remind everyone of the well found of this application and that, in this context, it is important to be able to determine the fluidity state of any membrane. Nevertheless, we consider that it is necessary to present briefly some reflexions on the biological meaning of the

B. Catoire (Editor)
Electron Spin Resonance (ESR) Applications
in Organic and Bioorganic Materials

membrane fluidity before dealing with one of the methods which allows to acceed to it.

For that, we need to see again the definition of the fluidity the way it is expressed in the dictionnary. "Fluid is said about body which without an own shape, can be easily deformed". That involves a kind of concordance between the different scales. These scales are respectively the spacial scale of the system, the temporal scale concerning the time needed to a particule to fill a volume egal to its free volume, and the interactional scale linked to the intensity and to the homogeneity of the interactions refasting these particules to each other. Thus, at the scale of a water glass, water is fluid because its molecules can move very rapidly each one in relation to the other. The weakness and the homogeneity of the hydrogen bonds allow them. So it is, at the scale of a mountain and for a considered time of several centuries, a glacier is fluid. On the contrary, a system is rigid if it includes too strong interactions for the spatio-temporal scale considered. It is simply rigid if those interactions remain isotropic. It becomes rigid and ordered if those last are anisotropic and belong to a field of force vectors (case of a balls flow for example, which have stabilized in a well ordered system under the influence of the gravitational interaction, that is to say of their weights).

Thus, a system is "fluid" if it presents a spatio-temporal and interactional concordance and if it is on another hand, caracterized by a certain homogeneity and isotropy. But the biological membranes are none homogen nor isotropic. So we cannot speak, rigorously, of membrane fluidity. Nevertheless it is possible to restrict the notion of fluidity to those of the lipidic phase. In fact, lipids are the more numerous molecules of the membranes. Moreover they are relatively homogene, as they are in majority of a phospholipidic nature. Their interactions almost exclusively of a Van der Waals type, are enough weak to allow molecular motions of above 10^{-8} to 10^{-4}s. Nevertheless, it exists an anisotropy of distribution of the phospholipids, the hydrophilic polar heads remain in surface and the paraffinic chains get settled in a more or less parallel way in the hydrophobic core. If the anisotropy is very marked, the way it is in the superficial region, it is better to use the term order (at its cristallographic sense) than rigidity. In the deep hydrophobic core, the anisotropy loss will be expressed in term of molecular motions, that is to say of fluidity.

Methods allowing to test this lipidic fluidity are mostly spectroscopic methods whose temporal windows of measure are in the same scale of sizes as molecular times (NRM, ESR, Fluorescence).

II - SPIN LABELLING

The electron spin resonance (ESR) is a spectroscopic method that studies the magnetic levels of electronic spin. This spin is observable when the spin number of the chemical specie is different from zero.

Five cases can be distinguished :

- The ions of transition metals that are often in a state in which their electrons are not paired (Fe^{3+}, Cu^{2+}, Mo^{5+}...).
- The compounds brought to the triplet metatable state.
- The paramagnetic centers created by a deficiency in the cristals.
- The electrons of conduction in the metals.

- The free radicals : most of them, frequently obtained in the course of oxydation-reduction reactions, are very reactive species and they have a short life time. Nevertheless, those of the nitroxide family are on the contrary extremely stable in the usual conditions of temperature and pressure. They slighthy disturb the biological environments as membranes and they have physical properties allowing a great facility for detection and a great sensitivity to structural variations of their environment. The spin labelling technics is based on these compounds which have all the required qualities to be good probes. In a simplified way, the hamiltonien describing the different energetic states in the free radical is :

$$\mathcal{H} = \beta_e Bg_e S + \beta_n Bg_n I + STI + aSI$$

The first term represents the interaction of the magnetic field with the electronic spin. Although g_e (Lande factor) is a tensor, its value remains at approximatively 2 for free radicals. For a nitroxide g_{ex} = 2.0098, g_{ey} = 2.0061, g_{ez} = 2.0027. The second term describes the field interactions with the nuclear spin. This one is important in NMR, but non existent in ESR because the used frequency is too much high to oberve the transitions at the nuclear spin level. The two last terms correspond to electronic spin and nuclear spin-interactions. The STI term describes the interactions between two magnetic moments, possessing a precise orientation in space. This kind of interaction is anisotropic and T is a tensor. These couplings are observed in free radicals with electronic orbital strongly dissymetric like nitroxides and also when the molecular movements are enough slow down in order not to cancell the orientation effects by average effect. The term aSI describes the Fermi isotropic coupling which is independent of the free radical movements and of its orientation in the magnetic field.

III - SPECTRA OF THE SPIN LABELLING

The interpretation of spin labelling spectra is based essentially on the dipolar coupling term STI. We can see (Fig.1) that the single electron of the nitroxide radical is mainly localized on the nitrogen atom, the spin density is about five times higher in the 2PZ orbital (perpendicular to the molecular plan) than on the 2PX and 2PY orbitals. When the magnetic field (H) which brings up the degeneration of electronic spin levels is parallel to the 2PZ orbital, the magnetic coupling with the

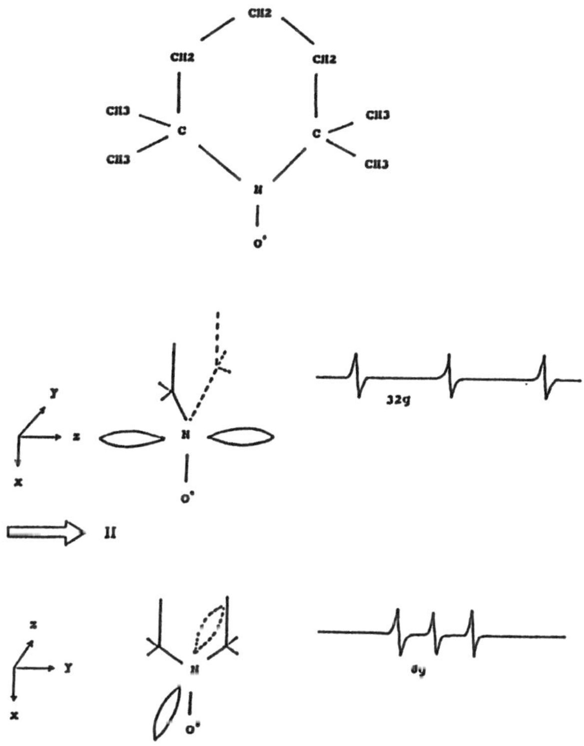

FIGURE 1 : Chemical formula of the tetramethyl piperidino-oxyl. The magnetic coupling with the nitrogen atom varies with the orientation of the magnetic field.

nitrogen atom (I=1) is maximum and the spectra are made of 3 lines (I = -1,0, -1), 32 gauss a part. In the perpendicular plan, the 3 lines are only 3 gauss a part.

A nitroxide in solution in a solvent of low viscosity presents a very rapid rotational motion whose correlation time τ_c is given by the Debye relation :

$$\tau_c = \frac{4 \pi n a^3}{3 kT}$$

in which (a) is the medium radius of the molecule and (n) the viscosity of the solvent. In those conditions, the rotational frequency is higher to that corresponding to the difference between the anisotropic coupling constants $a_z - a_x = 26$ gauss (a frequency equivalent to 73MHz in a field of 3500 gauss). Then we observe an anisotropic coupling $a = 1/3\ (a_x + a_y + a_z)$ 15 gauss.

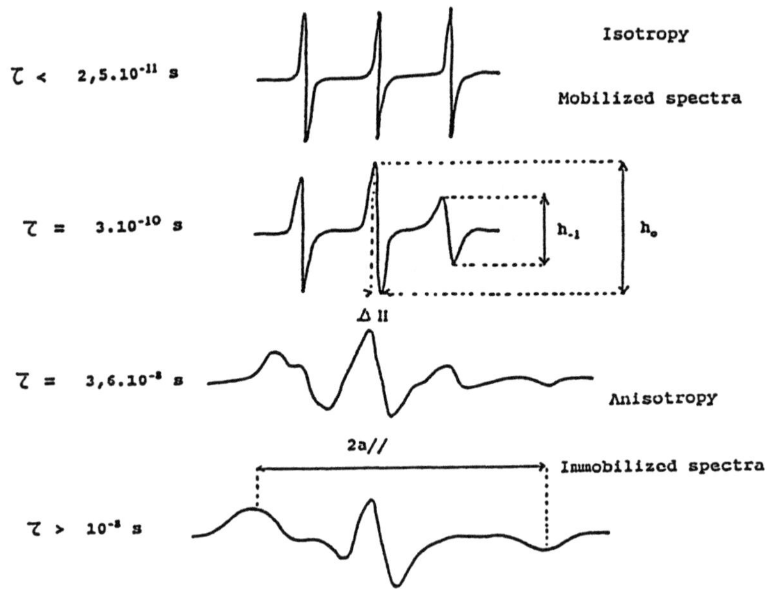

FIGURE 2 : Modification of the spectra anisotropy with the rotation frequency of the nitroxide radical.

If viscosity increases there is an anisotropic effect caracterized first by a spectral deformation with in a major part a widening of the high field line. As long as τ_c is smaller than 10^{-8}s wa can easily evaluate the correlation time by the following equation :

$$\tau_c = 6{,}5.10^{-10}.\,\Delta H\ \left(\sqrt{\frac{h_o}{h_{-1}}} - 1\right)$$

where (H), (h_o) and (h_{-1}) are defined figure 2 for isotropic or mobile spectra. For slower movements, higher than 10^{-8}s, interpretation is based on the distance 2a// between spectra externe lines, considerably extended to every field.

1 - Application to the proteins

The number of applications of this technic is considerable in the protein study field. We can easily fixe a nitroxide radical on reactive amino acid residues like thiol or amine fonctions. Then the spectroscopic response of the label depends on its localisation in the tertiary proteic structure. When fixed to the periphery, it will present few steric constraints and its spectrum will be "mobile". On the contrary if it is hidded in amino acid chains, its spectrum will indicate a strong hindrance to the rotation. A lot of proteins have been studied this way (2,3). Recently it has been shown that spectrin, a specific protein of the internal side of the red blood cells membranes was very thermosensitive (the spectra are modifying considerably with the temperature). Above all, its conformation was modified under the effect of divalent cations like calcium and magnesium (4).

In the molecular pharmacology field Chignell's work is

worth indicating (5). By this technic, this author showed that the active site of the carbonic anhydrase was deeply burried inside a chap of 14 Å deep. Several nitroxide derivatives of sulfamid inhibitors of these enzyme were used by fluctuating the lengh of the chain separating the sulfamide inhibitor grouping from the nitroxide label one. As the length was enough, the nitroxide was outside the chap and caracterised by the change of a spectrum strongly anisotropic which became isotropic.

2 - Application to the membranes

It is probably in the study of biological membranes that spin labelling provided the most of informations. In this method, nitroxide derivatives of fatty acids or phospholipids are inserted parallely to the paraffinic chains of the natural phospholipids of the membranes. These probes are structured in such a way that the principal orbital 2PZ of the single electron of nitroxide is parallel to the great axis of the carbon chain. In such a membrane system, we have to do the distinction between several situations refering to the orientation of the membranes and the degree of desorganization of fatty acid chains.

a) The membrane system is oriented and fatty acid chains are organized in a parallel direction to each other (saturated chains). According to the sample orientation in the magnetic field, we can observe a maximum SI coupling when the great axis of the fatty acids is parallel to the field (2T// or 2a//) and minimum on the contrary ($2T\perp$).

b) If membrane system is not directed and fatty chains are disorganized we can observe a spectrum of isotropic or mobil type.

c) When the membrane system is previously oriented but with disorganizated fatty acid chains or animed with a great rotation frequency, here again the spectra are of isotropoic type.

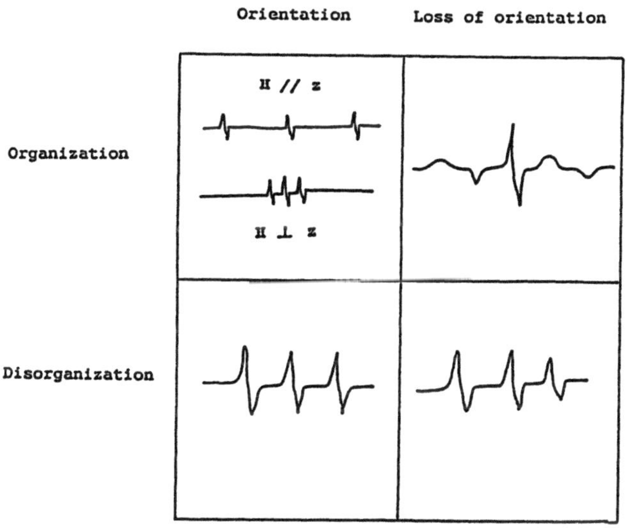

FIGURE 3 : Different types of spectra observed on a membrane system according to their orientation with regard to the magnetic field and the organization more or less parallel of these phospholipidic fatty chains.

d) At last, when fatty chains are organized and the membrane system is not oriented, the observed spectra are of anisotropic type. They should be considered as the sum of all the anisotropic elementary spectra corresponding to the different orientations of the chains in relation to the field. Thus we can define a parameter (S) which is the ratio of the measured constants to those of the maxima constants observed in a perfectly organized system.

$$S = \frac{2\,T//_o \quad - \quad 2T\perp_o \text{ (measured values)}}{2\,T//th \quad - \quad 2T\perp th \quad \text{(theoretical values)}}$$

This parameter is linked to the standard angular deviation (α) that the fatty chain present in relation with an axis perpendicular to the membrane plan.

$$S = 1/2 \quad (3 \cos^2 \alpha - 1)$$

In a suspension of biological membranes which are not oriented in the magnetic field, all of intermediary spectra can be shown. Their isotropic or anisotropic features are depending on the disorganization or the disorganizability (notion close to the fluidity) or the fatty acid chains. The applications of this method on the studies of membranes are very numerous and it would be impossible to cite them in the context of this report. Nevertheless we can classify them in the following headings.

- Fondamental studies of the membrane structure

The spin labelling method allows to confirm the double layer structural state of the membrane phospholipids. This method has shown up a "gradient of fluidity" in every double layer. The more or less deep labelling of these layers has shown a high parallelism of the fatty chains in the polar heads region and an increasing disorganization and disorganizability when shifting near the hydrophobic core (6). It is also thanks to the spin labelling that the previous measures of the lateral diffusion coefficients and the flip-flop times of the phospholipids have been made (7,8). More recently, this method has been used to characterize the asymetrical repartition of the membrane phospholipids (9).

The effect of the chain length and the insaturation degree of fatty acids as well as the rate of cholesterol have been clearly studied. More generally, numerous membrane systems have been compared in order to establish correlations between their physico-chemical structure, their biochemical composition and their biological functions (10).

In this context of triple relation composition-fluidity-function, the spin labelling is usefull to caracterize the estimation of global fluidity determined at a given temperature. The results are more rarely expressed in fluidity gradients as a function of the layer depth. This method has also allowed dynamical and theoretical approaches as the phenomemum of hydrophilic-hydrophobic balance (11) or the notion of structural attractors (12). In this last notion, the membrane disorganizability is linked to the breaking down of the different types of bindings between the membranes constituents and it can be apprehended by the determination of the thermic susceptibility of the membrane. This last, like the activation energy activation can be determined by the slope of the curve describing the rotational frequency as a function of temperature.

- Applied studies

It is certainly in the pharmacological field that the spin labelling and the membrane fluidity have found the highest number of applications. Because of the restrictions already formulated concerning the fluidity and the anisotropy of the membranes, spin labelling cannot give absolute values of membrane fluidity. In return, this method is very sensitive and it gives rise to the comparizon of two membranes from the same origine but with one of them submitted to a pharmacological effector. So, structural effects of the local anesthetics and the detergents on membranes have been for the more studied. A lot of works have been also made on the phenothiazines and, it particular, on the chlorpromazine. The spin labelling associated with the photochemistry has permitted to show up a rigidifiant action of these drugs in the polar area of the membrane (13). This pharmacological action of chlorpromazine is competitive with the physiological one of calcium, it is polyphasic and unfolds synchronously with the membrane structure and the adenylcyclasic activity (14). It can be found in the litterature numerous exemples of pharmacological compounds interacting with membranes studied with the spin labelling method. Amoung them, ajoene, extracted from garlic with antithrombotic properties which has the particularity of changing the hydrophobic structure, but not the hydrophilic one.

Numerous studies have been also applied in the field of membrane pathology. So appeared the idea that some myopathies would be generalized membrane disorders and that thanks to pointing out fluidity modifications in the erythrocyte membrane (15,16). Others hereditary disorders are characterized by this type of alteration : this is the case for the Hutington disease (17). Modifications of myeline structure for mouse having development changes of the nervous system have been also studied with this method (18,19). Precocius changes have been revealed for the synaptic membrane of the mouse in the scrapie disease evolution. (20). Changes of the erythrocyte membrane structure have been revealed in course of the blood preservation and connected to its rheological properties (21). More recently a modification of the activation energy of fatty acids has been revealed in the erythrocyte membrane of rats having a high blood preasure from genetic origin (22).

After all, studied by spin labelling have been made in the intestinal brush border membranes on animals (23), with deficiency of essential fatty acids and on blood platelets of subjects attached by troubles of coagulation (24).

IV - CONCLUSIONS

Indeed, the spin labelling method is only one of the possibilities of utilisation of the ESR in biology. This method is neither new nor very spectacular. Nevertheless it remains the reference method of the structural study of the biological membranes in their whole. It has often been used to evaluate the "global membrane fluidity". But we have indicated that this term has to be considered very cautiously. In fact, and taking into account the temperature variations in order to obtain an Arrhenius representation this method allows to appreciate with a high sensitiveness the influence of a lot of membrane effectors on the disorganizability of the membrane phospholipidic phase. The thermic sensitiveness of a membrane seems to be a parameter more significant than its fluidity state because it is directly binded to the interactions forces that are linking the membrane constituents among themselves. Now, the true relation between the structure and the function of biological phenomena of the biomembranes consists probably in these constituents organization.

R E F E R E N C E S

-=-=-=-=-

1 - SINGER S.J. and NICOLSON G.L. (1972)
The fluid mosaic model of the structure of cell membrane.
Science, N.Y., 175, 720-731.

2 - OHNISHI S., BOEYENS J.C.A. and Mc CONNELL (1966)
Spin labeled hemoglobin crystals.
Proc.Nat.Acad.Sc.U.S., 56, 809-813.

3 - JOHNSON M.E. (1978)
Librational motion of an "Immobilized" spin label.
Hemoglobin spin labeled by a maleimide derivative.
Biochemistry, 17, 1223-1228.

4 - CASSOLY R., DAVELOOSE D. and F.LETERRIER F. (1980)
Spin labeling of human spectrin. Effects of temperature, divalent cations and reassociation with erythrocyte membrane.
Biochim.Biophys.Acta., 601, 478-489.

5 - CHIGNELL C.F. and STARKWEATHER D.K. (1973)
A spin label study of horse erythrocyte carbonic anydrases C and D.
Comp.Biochem.Physiol., 46 B, 813-822.

6 - BERLINER L.J. (1976)
Spin labeling, Theory and Applications. Ac.PRESS.N.Y.

7 - DEVAUX P. and Mc.CONNELL H.M. (1972)
Lateral diffusion in spin labeled phosphatidylcholine multibilayers.
J.Am.Chem.Soc., 94, 4475-4481.

8 - ROUSSELET A., GUTHMANN C., MATRICON J., BIENVENUE A. and DEVAUX P. (1976)
Study of the transverse diffusion of spin labeled phospholipids in biological membranes. Human red blood cells.
Biochim.Biophys.Acta., 426, 357-384.

9 - SEIGNEURET M. and DEVAUX P. (1984)
ATP dependent asymmetric distributions of spin labeled phospholipids in the erythrocyte membrane : relation to shape changes.
Proc.Nat.Acad.Sci.USA, 81, 3751-3755.

10 - VIRET J. and LETERRIER F. (1976)
A spin label study of rat brain membranes effects of temperature and divalent cations.
Biochim.Biophys.Acta., 436, 811-824.

11 - SHINITZKY M. (1984)
Membrane fluidity and receptor function.
Biomembranes, 12, 585-601.

12 - VIRET J. and DAVELOOSE D. (1989)
Biophysical interpretation of membrane fluidity by catastrophe theory.
J.Theor.Biol., 140, 51-82.

13 - LETERRIER F., MENDYK A. and VIRET J. (1976)
Interaction of chlorpromazine with biological membranes. A photochemical study using spin labels.
Biochem.Pharmacol., 25, 2469-2474.

14 - SALESSE R., GARNIER P., LETERRIER F., DAVELOOSE D. and VIRET J. (1982)
Modulation of adenylate cyclase activity by the physical state of pigeon erythrocyte membrane : parallel drug-induced changes in the bilayer fluidity and adenylate cyclase activity.
Biochemistry, 21, 1581-1586.

15 - BUTTERFIELD D.A., CHESNUT D.E., APPEL S.H. and ROSES A.D. (1976)
Spin label study of erythrocyte membrane fluidity in myotonic and Duchenne Muscular Dystrophy and congenital myotonia.
Nature, 263, 159-161.

16 - LAURENT M., DAVELOOSE D., LETERRIER F., FISCHER S. and SCHAPIRA G. (1980)
A spin label study of the erythrocyte membrane in Duchenne Muscular Dystrophy.
Clin.Chim.Acta., 105, 183-194

17 - BUTTERFIELD D.A., OESWEIN J.Q. and MARKESBERY W.R. (1977)
Electron spin resonance study of membrane protein alterations in erythrocytes in Huntington's disease.
Nature, 267, 453-455.

18 - LETERRIER F., BRETON J., DAVELOOSE D., VIRET J., LE SAUX F. and POLLET S. (1978)
Physico-chemical state and calcium reactivity of normal and quaking mouse myelin.
Biochim.Biophys.Acta., 507, 525-530.

19 - VIRET J., LETERRIER F. and BOURRE J.M. (1979)
A spin label study of sciatic nerves from Quaking, Jimpy and Trembler mice.
Biochim.Biophys.Acta., 558, 141-146.

20 - VIRET J., DORMONT D., COURT L., LETERRIER F., GIBBS and GAJDUZCEK (1981)
Structural modifications of nerve membranes during experimental scrapie evolution in mouse.
Biochem.Biophys.Res.Commun., 101, 830-836.

21 - ESSABBAH H., LACOMBE C., FABRE G., SAINT-BLANCARD J., DAVELOOSE D., MOLLE D. et LETERRIER F. (1981)
Etude rhéologique en régime transitoire de sang conservé à l'état liquide.
ReV.Franc.Transf.Immuno-hématol., 4, 357-373.

22 - DAVELOOSE D., VIRET J., MOLLE D. and LETERRIER F. (1980)
Mise en évidence par marquage de spin d'une modification structurale de la membrane erythrocytaire du rat génétiquement hypertendu.
C.R.Acad.Sc., 290, 85-88.

Potentiel Transmembranaire dans les Vésicules Artificelles. Fragmentation des Vésicules. Encapsulation

J.Faure, R.Pansu, J.Roncin, L.Liu
Physicochimie des Rayonnements
Bat 350 Université PARIS SUD UA 75 du CNRS
F-91405 Orsay CEDEX

Introduction

Les milieux microhétérogènes et en particulier les vésicules sont des milieux attrayants pour l'étude et l'amélioration des réactions chimiques. Par exemples on peut envisager de mettre à profit le champ électrique membranaire (Figure1) pour favoriser la séparation de charge photo-induites au cœur d'une membrane. Dans ce but nous avons étudié des vésicules synthétiques chargées en milieu de faible force ionique, et plus particuliairement les vésicules de DODAC (Chlorure de Dioctadecyldimethylammonium) et de DHP (Acide Dihexadecylphophorique) préparées par sonication.

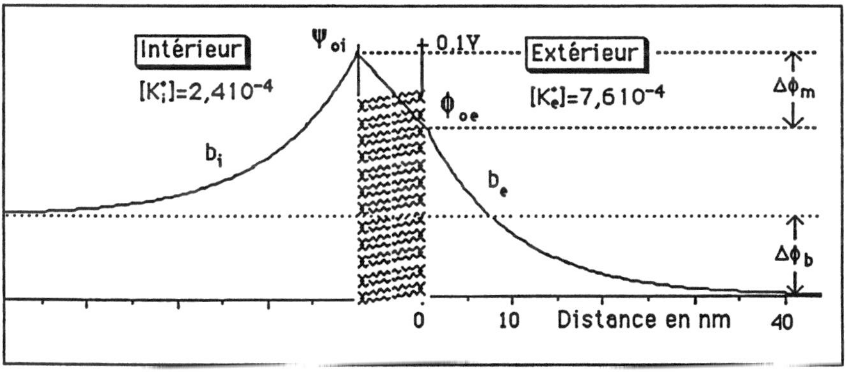

Figure 1 : Ce schéma présente un exemple de la variation du potentiel électrique au voisinage d'une bicouche chargée. On reconnait dans la phase aqueuse, de part et d'autre de la membrane, la double couche de Gouy-Chapman où il existe un champ électrique dû à l'éloignement des contre-ions chargés par rapport aux têtes polaires de la membrane. Le champ membranaire et

B. Catoire (Editor)
Electron Spin Resonance (ESR) Applications in Organic and Bioorganic Materials

Elles sont accompagnée des valeurs de différence de potentiel transmembranaire et de fraction de vésicules que l'on peut en déduire de ces mesure. Dans le cas du DODAC comme du DHP, on a pu mesurer le champ électrique membranaire. Le signe mesuré est en accord avec les prévisions. Mais l'originalité principale des ces études et la fraction très importante de tensioactif sous forme de fragments ouverts. De tels fragments n'ont étés observés que par microscopie electronique par Kunitake[2]. Les mesures d'encapsulation montrent que les préparations de vésicules chargées contiennent des objets ouverts.

Les études de diffusion de la lumière montrent que ces fragments sont petits. La proportion de fragment dépend de la durée de la sonication. Les vésicules sont fragmentées par la cavitation produite par les ultrasons. La proportion de fragments dépend aussi de la charge superficielle portée par la membrane. Que se soit par neutralisation des charge des têtes polaires, par augmentation de la force ionique ou par chauffage, les fragments disparaissent. Les suspensions de fragments ont dans notre cas une stabilité de type colloïdale. La disparition des fragments s'accompagne de la formation d'une population de vésicules monodisperse en taille. L'existence de fragments de membranes comme intermédiaires dans la formation des vésicules a été postulé pour rendre compte de la monodispersité en taille de ces objets[3] . Nos études constituent la première démonstration expérimentale de la validité de ces modèles.

Études des fragments

Nous avons étudié ces nouveaux objets : s'agit il de micelles (liquide) ou de fragments de membrane (présentant une transition de phase liquide/cristal caractéristique de nombreuses bicouches)? En effet la transition de phase est un phénomène coopératif. Si la phase devient trop petite ces phénomènes coopératifs peuvent disparaître. Nous avons de nouveau fait appel à la RPE.

La transition de phase s'accompagne d'une discontinuité dans la viscosité membranaire et/ou d'un changement dans le mouvement des sonde dissoute dans la membrane. Ces changements se reflètent dans les spectres. Nous avons reporté sur la figure 4 le rapport de la raie à champ fort a la raie centrale du nitroxyde 16NS pour deux échantillons : des fragments de bicouche obtenus par sonication du DODAC et des multicouches de DODAC. La transition de phase dans les multicouches se traduit par une discontinuité qui ne se retrouve pas dans l'échantillon de fragments. La diminution de la taille amène une disparition de la transition de phase. Les disques sont la différence de potentiel transmembranaire sont dus au passage d'ions à travers la membrane.

La RPE a joué un rôle déterminant dans cette étude. Nous avons ainsi mesuré pour la première fois la différence de potentiel transmembranaire due à des interactions de contacte prévues par la théorie. Nous avons également constaté par les mesures d'encapsulation de sondes paramagnétiques qu'il était possible de détruire les vésicules par sonication, de produire et de stabiliser des fragments de membranes ouverts. Ceci nous a conduit à revoir les modèles de formation des vésicules. Nous avons ensuite identifié les fragments produits par la sonication comme des micelles grâce à l'étude des mouvements de sondes nitroxydes dans ces objets.

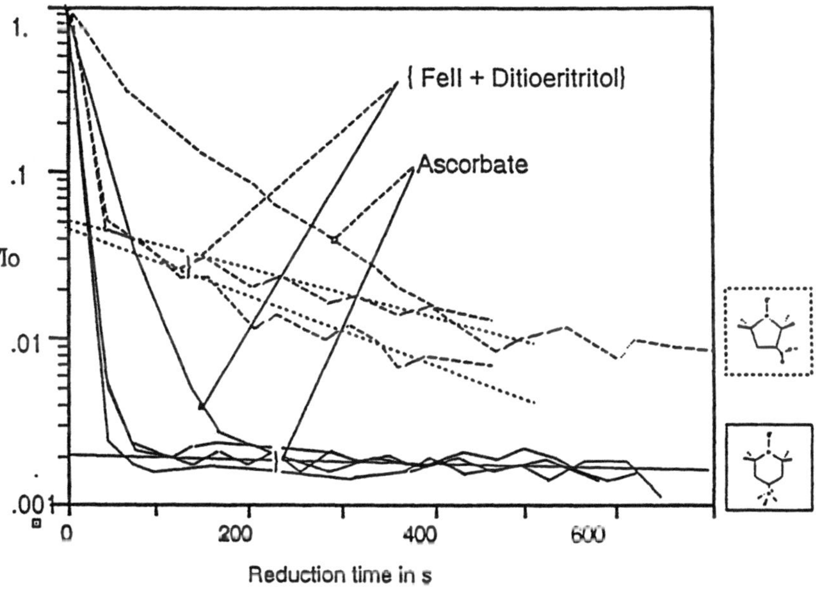

Figure 2 : Cette figure présente une mesure d'encapsulation. L'intensité du signal RPE est représentée en fonction du temps après l'ajout du réducteur. Les résultats obtenus aux moyen de deux réducteurs différents sont comparés.

Recherche de sondes membranaires

Nous avons utilisé des sondes nitroxydes. Mais nous cherchions des sondes susceptibles de se dissoudre spécifiquement à l'interface. La localisation des sondes n'était pas connue dans le cas des tensioactifs synthétiques que nous étudions. Cependant l'étude de la forme des spectres permet d'identifier les sites. Nous avons constaté que des deux interactions intervenant dans l'affinité d'une sonde pour un site, les interactions électrostatiques sont plus fortes que les interactions hydrophobes. Par exemple la sonde cationique la plus hydrophobe disponible, le CAT16, ne s'absorbe que partiellement sur les surfaces cationiques. La mesure du coefficient de partage est rendu possible grâce à la grande différence de spectre des forme libre et adsorbée, illustrée sur la figure 3. Ceci nous a permis de réaliser des mesures d'encapsulation avec cette molécule.

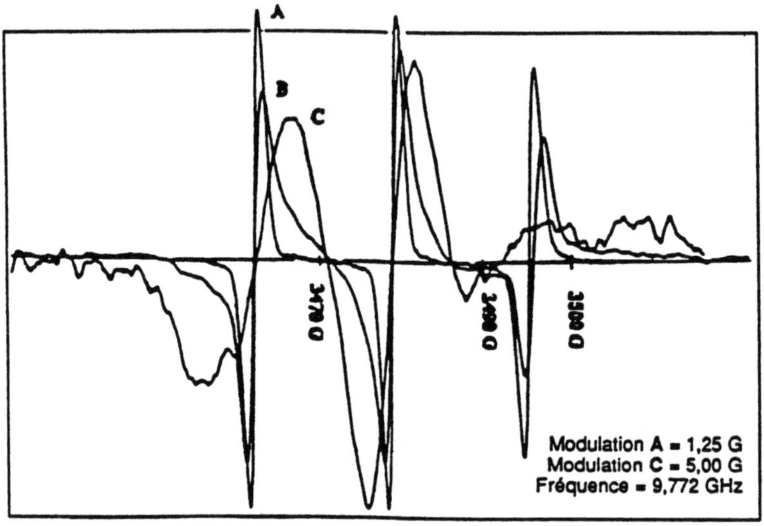

Figure 3 : Trois spectres du CAT16 sont présentés pour illustrer l'influence du site de solubilisation sur la forme du spectre. A - Signal obtenu pour une sonde en phase aqueuse; B - Signal obtenu en présence de DODAC 0,1mol/l; C - Signal obtenu après réduction (molécules essen tiellement absorbées dans la membrane)

L'encapsulation

La propriété la plus marquante des vésicules est l'encapsulation, les vésicules possédant une phase aqueuse interne isolée du reste de la solution qui peut piéger des solutés. La mesure de l'encapsulation ζ est le rapport entre les molécule situées à l'intérieur des vésicules au nombre total de molécules de la préparation.

Suivant la nature et la localisation des sondes ceci permet de mesurer la taille, la concentration de vésicules, la contamination par des structures ouvertes et les différences de potentiels transmembranaires. Pour cela de nombreuses méthodes ont été proposées. La RPE utilisant des sondes paramagnétiques s'avère la plus intéressante car la méthode est sensible (nous avons utilisé des concentration de l'ordre de 10^{-4} à 10^{-3} mol) et présente une grande dynamique de mesure (illustrée sur la figure 2). L'étude de la forme des spectres des sondes permet de plus de diagnostiquer et de traiter de manière fiable les problèmes à l'existence de nombreux sites de solubilisation dans les systèmes microhétérogènes.

Résultats

La mesure de la différence de potentiel transmembranaire et de la fraction de vésicule se ramène à l'étude de l'encapsulation de sonde membranaires chargées. En effet, en première approximation, l'encapsulation est reliée à ces deux grandeurs par :

$$\frac{\zeta_+}{\zeta_-} = f \exp(2\frac{\Delta\phi}{kT})$$

où f est la fraction de tensioactifs organisée en vésicules,

$\Delta\phi$ est la différence de potentiel transmembranaire,

ζ_+ et ζ_- sont respectivement les encapsulations par des sondes cationiques et anioniques, le rapport molaire des de la population protégée à la population totale.

Le volume intérieur et la différence de potentiel entre coeur de phases ont été mesurés au moyen de sondes solubilisées dans la phase aqueuse

Pour distinguer l'intérieur de l'extérieur dans les mesures d'encapsulation plusieurs techniques ont été proposées. La chromatographie est la plus courante. La chromatographie de perméation de gel permet de retirer les solutés de la phase externe. Cependant on observe que les vésicules peuvent être détruites. Une autre technique, propre aux sondes paramagnétiques, est de rajouter dans la phase extérieure un ion tel que $Fe(CN)_6^{4-}$ (à une concentration de 0,1mol). L'interaction d'échange entre les nitroxydes et l'ion paramagnétique élargit les spectres des sondes extérieures. Celle-ci disparaissent virtuellement du spectre, permettant la mesure de la population intérieure qui, elle, est protégée. Du fait de la grande sensibilité à la force ionique de nos échantillons nous avons préféré mesurer la population intérieure par réduction de la population extérieure.

Un exemple de réduction est donné par la figure 2. L'intensité du signal RPE, mesurée pic à pic, est représentée en fonction du temps après l'ajout du réducteur. Après un déclin rapide correspondant à la réduction de la population extérieure une composante lente apparait correspondant à la population intérieure. La vitesse de déclin de la composante lente dépendant de la vitesse de diffusion des réactifs à travers la membrane et de la vitesse de destruction des vésicules on a ainsi un moyen de contrôler extemporanément la stabilité de l'échantillon. De la dizaine de réducteurs criblés l'ascorbate[1] et le mélange Fe^{2+} DTT (mélange qui à notre connaissance a été utilisé pour la première fois) se sont révélés les plus appropriés.

Discussion

Des valeurs typiques d'encapsulation sont présentées dans le tableau 1.

DODAC	Encapsulation		
5 NS	$8{,}3 \pm 0{,}5 \; 10^{-2}$	+ 12 mV	Transmembrane Potential
16 NS	$5{,}3 \pm 5 \; 10^{-2}$	$13 \pm 3 \; 10^{-2}$	Vesicle Fraction
DHP			
5 NS	$35 \; 10^{-2}$	- 6 mV	Transmembrane Potential
16 NS	$58 \; 10^{-2}$	$90 \; 10^{-2}$	Vesicle Fraction

Table 1 : L'échantillon de DODAC est totalement ionisé. L'échantillon de DHP ne l'est que partiellement, l'essentiel des têtes polaires étant sous la forme acide neutre.

en dessous de la dimension caractéristique permettant de parler de phase.

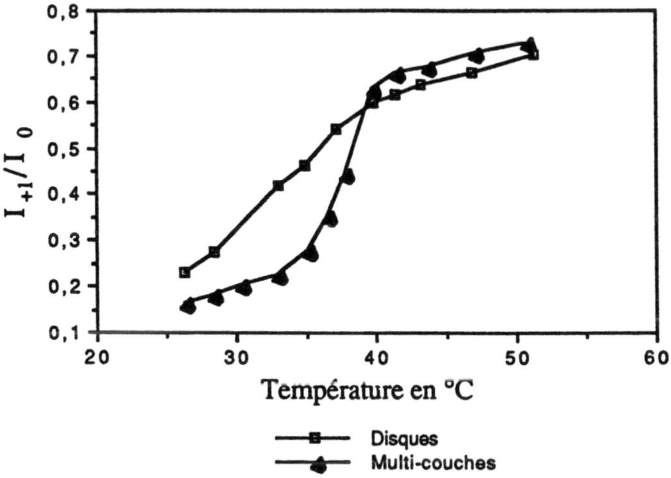

Figure 4 : La fluidité membranaire dans des échantillons de DODAC est représentée en fonction de la température. La disparition de la discontinuité observée dans les multi-couches indique la disparition de la transition de phase.

Conclusion

Nous avons mis en évidence l'existence de fragments de membranes dans les suspensions de tensio actifs chargés obtenus par sonication. Les fragments de membranes que nous obtenons sont des objets nouveaux qui permettent d'explorer l'apparition des phénomène coopératifs dans les petits assemblages moléculaires.

---

1 Lim, Y.Y. & Fendler, J.H. J.Am.Chem.Soc. 101, 4023-9 (1979).

2 Kunitake, T. & Okahata, Y. J.Am.Chem.Soc. 99, 3860-1 (1977).

3 Lasic, D.D. La Recherche 212, 905-13 (1989)

Electron Spin Resonance as a Tool for the Study of Langmuir-Blodgett Films of Nitroxides and Metal Alcanoates

Giacomo Martini, Francesca Bonosi, Gaetano Carubia, M.Francesca Ottaviani, and Gabriella Gabrielli

Department of Chemistry, University of Florence,
Via G. Capponi 9 - 50121 Firenze, Italy.

Abstract

The application of the ESR to the study of the Langmuir-Blodgett films is discussed by reference to copper stearate films (as an example of material with peculiar magnetic properties) and to films of nitroxide derivatives of stearic acid (as very simple examples of membrane mimetic). Particular emphasis is given to the need of a preliminary, detailed study on the film-forming monolayers, whose physical and structural properties determine both reproducibility and homogeneity of the film. The different condensation of the copper stearate monolayers in the presence and in the absence of a condensing ion such as Ba^{2+} has allowed us to recognize that the formation of a Cu-Cu dimer at the water/air interface is the kinetic parameter which determines the film stability. The spin triplet-

B. Catoire (Editor)
Electron Spin Resonance (ESR) Applications
in Organic and Bioorganic Materials

doublet state of the film has been investigated by ESR of the S=1/2 monomer adduct formed after treatment of [$Cu_2(St)_4$] with ammonia vapours. On the other hand, the ESR signals obtained from LB films of pure nitroxides or from nitroxides diluted in stearic acid have given details on the spin-spin interactions and on the shape anisotropy. The motional parameters evaluated from the lineshape are of the correct order of magnitude expected for multilayer systems such as the membranes.

INTRODUCTION

Langmuir-Blodgett (LB) films are ultrathin, molecularly ordered assemblies with thickness variable from a few nm (mono- or bilayers) to ~100 nm (multilayers). They have relevant interest in both technological and scientific field because of their use as electronic and electric devices, sensors, transductors, magnetic systems, and as very simple model biomembranes [1-3]. The increasing attention devoted to these systems is certainly due to their uncommon physical and chemical properties, including their peculiar magnetic and electric properties because the LB films can be controlled on a molecular scale. The first example of a literally two-dimensional magnet is indeed a LB film of manganese stearate, whose structural and magnetic properties have been studied by Pomerantz and coworkers [4-6]. On the other hand, lamellar conduction LB films have been obtained by Barraud and coworkers [7-9] in two successive steps, namely by transference of a precursor salt of N-docosylpyridinium with tetracyanoquinodimethane onto a CaF_2 substrate and by *in situ* partial oxidation with iodine vapours. Electron Spin Resonance spectroscopy, together with UV-visible, IR spectra and linear dichroism, has been used to characterize these particular assemblies [9,10].

Similar systems also show the behaviour predicted by the random-exchange Heisenberg antiferro-magnetic chain (REHAC) model, as proved with ESR by Ikegami and coworkers [11-12]. Photoconductor and semiconductor LB films have also been studied with magnetic resonance techniques [13].

The ESR technique has also been widely employed in studies of LB films containing transition metal ions, particularly Cu^{2+} [15-19]. Extensive literature exists on LB films of copper compounds with phtalocyanines [15-17], porphirazines [18], and porphirines [19]. The dependence of the ESR spectra on the temperature and on the orientation of the film with respect to the applied magnetic field is generally used to determine the magnetic parameters and the structural order of the molecular films.

The LB films have also been recently built up with fatty acids and with their derivatives because these films are often considered the first step for a simple model of natural membranes [20,21]. They are therefore useful models for the study of several physical properties, such as lateral diffusion, phase transitions, etc.. Moreover, the addition of small amount of ESR-active components to diamagnetic LB films is the starting point for the study of the motional anisotropies of the hydrophobic units and of their order degree [22,23].

We have, however, to keep firmly in mind that before any analysis of the LB films having sense, a detailed characterization of the spreading monolayers must be carried out. This is a very general approach and it has been demonstrated recently in the case of metal alcanoates [24,25].

In this paper we report a short overview on two different systems for which the combined use of monolayer characterization and ESR spectroscopy allowed us to obtain both structural and dynamic information. This was done for LB films built up with fatty acid monolayers spread on aqueous subphases containing copper ions (and barium) and with pure and diluted nitroxides (with the doxyl moiety inserted in different positions of the stearic acid chain).

LB FILMS OF METAL ALCANOATES

As outlined in the introductory section the properties of the monolayer from which a LB film is to be formed condition not only the corresponding properties of the multilayer but also the possibility of transference and deposition. This is particularly true when LB films are formed from fatty acid monolayers on substrates containing metal ions [24]. The interactions between hydrated ions in the subphase and polar heads of the surface spread amphiphylic molecules must be characterized to find the best experimental conditions for the LB film formation.

Let us consider the case of copper stearate which gives homogeneous, reproducible films only when deposited as multilayers on a solid support starting from stearic acid monolayers spread on an aqueous subphase containing both Cu^{2+} and Ba^{2+} ions [24]. The formation of mixed monolayers of stearic acid bonded to Ba^{2+} and/or to Cu^{2+} has been proved through several techniques, including ESCA, TEM, SEM.

Two main problems remained to be solved:

i) Why pure copper stearate, as well as copper arachidate and copper behenate [25], does not give LB monolayers and multilayers, whereas LB films may be formed with manganese stearate or behenate and arachidate?

ii) Which are the magnetic properties of the $CuSt_2$ films as itself and in the presence of $BaSt_2$.

Both these questions may be solved by taking into account the ESR properties of Cu(II).

Before analyzing the ESR results from the $CuSt_2$ LB films a short summary must be given on the thermodynamics of the monolayers built up with the Ba/Cu stearate and on the magnetic properties of the simple copper alcanoates.

The spreading isotherms of the stearic acid are markedly influenced by the presence of metal ions in the aqueous subphase [24]. From the values of the compressibility moduli and of the limiting areas, an expansive effect on the monolayer is observed with Cu^{2+} ions in the subphase. The simultaneous occurrence of both $Cu^{2+}$and Ba^{2+} with $[Ba^{2+}]/[Cu^{2+}]$ ratios ranging from 0.18 to 15 leads to a progressive condensation of the film and this condensation is the more pronounced the higher $[Ba^{2+}]$ is. A satisfactory condition for monolayer transference was verified for $[Ba^{2+}]/[Cu^{2+}] = 15$. Figure 1 shows the spreading isotherms of this system at T = 288, 293, and 298 K. The practical temperature independence indicates that strong, attractive forces amidst the lateral hydrophobic chains are acting.

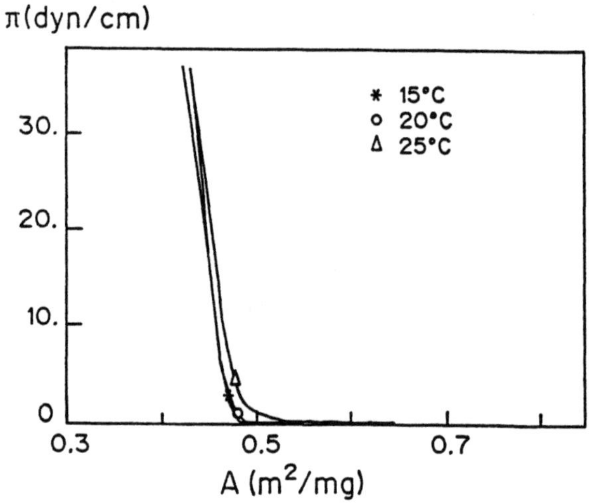

Figure 1 - Spreading isotherms of stearic acid on aqueous subphase containing $[Cu^{2+}] = 2x10^{-6}$ mol/L and $[Ba^{2+}] = 3x10^{-5}$ mol/L at pH 7.2.

At 293 K the compressibility modulus, C_s^{-1}, was 345-375 $dyn.cm^{-1}$ in the surface pressure range 8.2-37 $dyn.cm^{-1}$, with a limiting area, A_0, of 0.47 m^2/mg. This represents a sensibly more condensed phase (pseudo-solid phase) than in the absence of Ba^{2+} ions (C_s^{-1} = 88-107 $dyn.cm^{-1}$, A_0 = 0.65 m^2/mg).

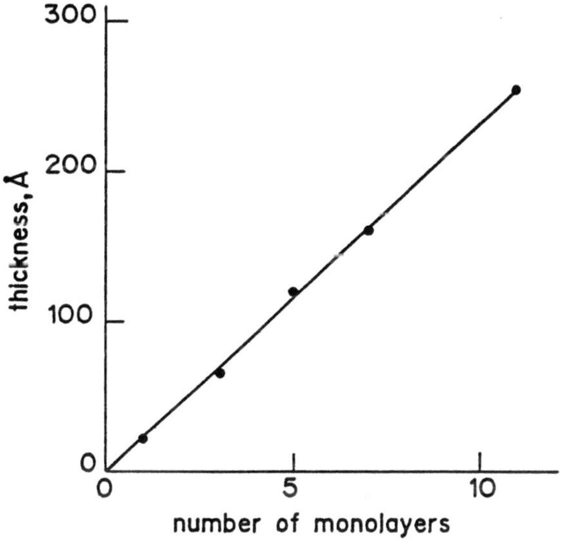

Figure 2 - Ellipsometric thickness of multilayer films of Cu/Ba stearate deposited on a chromium-coated quartz plate as a function of the number of layers. Same subphase as in fig. 1.

From ellipsometric measurements on mono- and multilayers deposited on chromium-coated quartz plates, the transference results neither quantitative nor reproducible when Cu^{2+} is the only metal ion in the subphase. Practically only one monolayer is transferred with a mean thickness of ~19 Å, which is lower than the thickness of a monolayer of pure stearic acid. This has also been found with copper behenate monolayers [25]. This situation is largely improved by the presence of Ba^{2+}. Figure 2 shows the thickness of a Cu/Ba stearate ([Ba]/[Cu]=15) film as a function of the number of transferred monolayers. An almost perfectly linear increase is observed, with an averaged thickness of 23 Å, well in line with the thickness of other metal stearates [6,26].

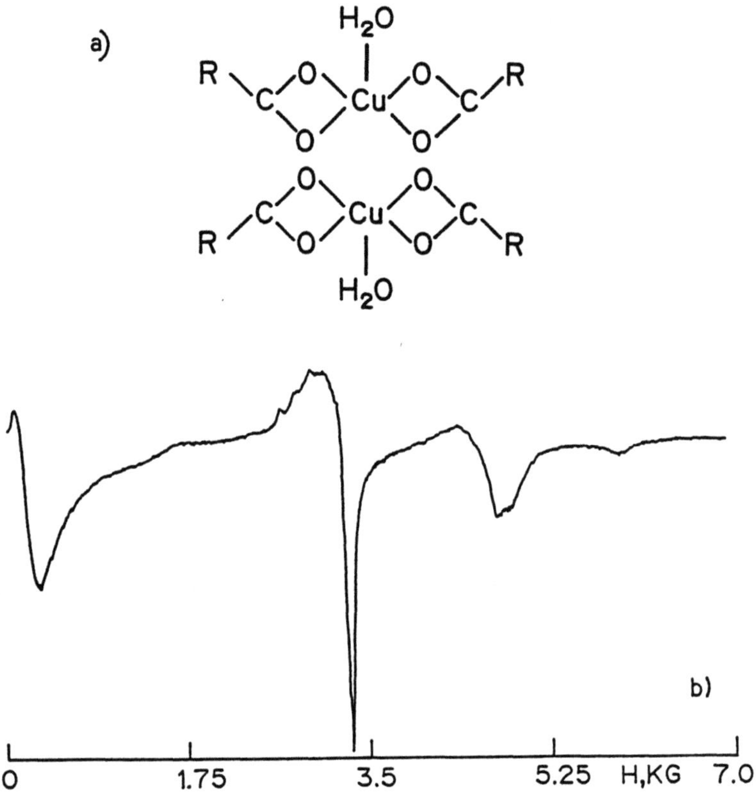

Figure 3 - Structure of dimeric copper alcanoates (a) and ESR spectrum of copper stearate (b; T = 115 K).

Figure 3a shows the general structure of the copper alcanoates [27] and figure 3b shows the 115 K ESR spectrum of $CuSt_2$ (powder). Most of the observed absorptions are typical of a dimeric (S = 1) copper complex [27-32]. The spectrum of the same sample at room temperature is sensibly broader. The absorption in the 2700 to 3300 G magnetic field range is due to isolated, monomer Cu^{2+} ions (S = 1/2) in almost axial symmetry ($g_{\parallel}$= 2.353, $g_{\perp}$ = 2.078, $A_{\parallel}$ = 143 G), present as an impurity in the bulk stearate. Similar monomer spectra have also been reported in dimeric copper acetate with various hydration degree [28], in anhydrous copper salts of monocarboxylic acids [29], in dimer copper complexes of linear-chain fatty diacids [30]. The other lines of the spectrum in fig. 3a, namely the absorptions at 203 G, 460 G, and 5912 G, correspond to B_{z1}, B_{z2}, and $B_{\perp 2}$ absorptions of dimeric copper systems [27,28,31,32]. They are described by the spin Hamiltonian [31]:

$$H = \beta B.g.S + D[S_z^2 - S(S + 1) + E(S_y^2 + S_x^2)] \qquad [1]$$

where D and E are the zero-field splitting parameters. From the theory of the triplet state ESR spectra [27,31], the following parameters are calculated for the $[CuSt_2]_2$ species:

$$g_{\parallel} = 2.35 \qquad g_{\perp} = 2.08 \qquad D = 0.34\ cm^{-1}$$

in a very good agreement with those of other copper alcanoates [31,33,34].

If collapsed monolayers of the Cu/Ba stearate are

analyzed through ESR, only a weak peak at g = 2.1 is observed at 298 K and it is due to $Cu(H_2O)_6^{2+}$ ions imbedded in the three-dimensional network of the disordered multilayer. No other absorptions are observed. Because collapsed materials arise from monolayer separation, dimeric Cu species are formed in the monolayer itself. The same lack of signals occurs with a LB film (50 layers) obtained from the same monolayer. The intensity of the triplet state lines is too low at room temperature to be detected even after several spectral accumulation.

Lack of ESR signals from Cu ions is also reported by Chollet [35] in LB films of copper dioctadecyl-dithiocarbamate (DDC-Cu) and by Messier and Marc [36] in a LB film of copper stearate. The latter system differs from ours because it was prepared by a different procedure and the Ba^{2+} have not been used to improve the layer deposition. In addition no thermodynamic characterization of the system is given.

According to the known reactivity of copper alcanoates with nitrogenous ligands in liquid solution [37,38] to give adducts that may be both dimeric and monomer adducts, ESR spectra from copper monomer units are easily obtained when a saturated dioxane solution of copper stearate is treated with gaseous ammonia. The micellar solution [39] turns from a pale blue to a violet colour and gives rise to a four-line ESR spectrum with $\langle g \rangle = 2.134$ and $\langle A_{Cu} \rangle = 62$ G.

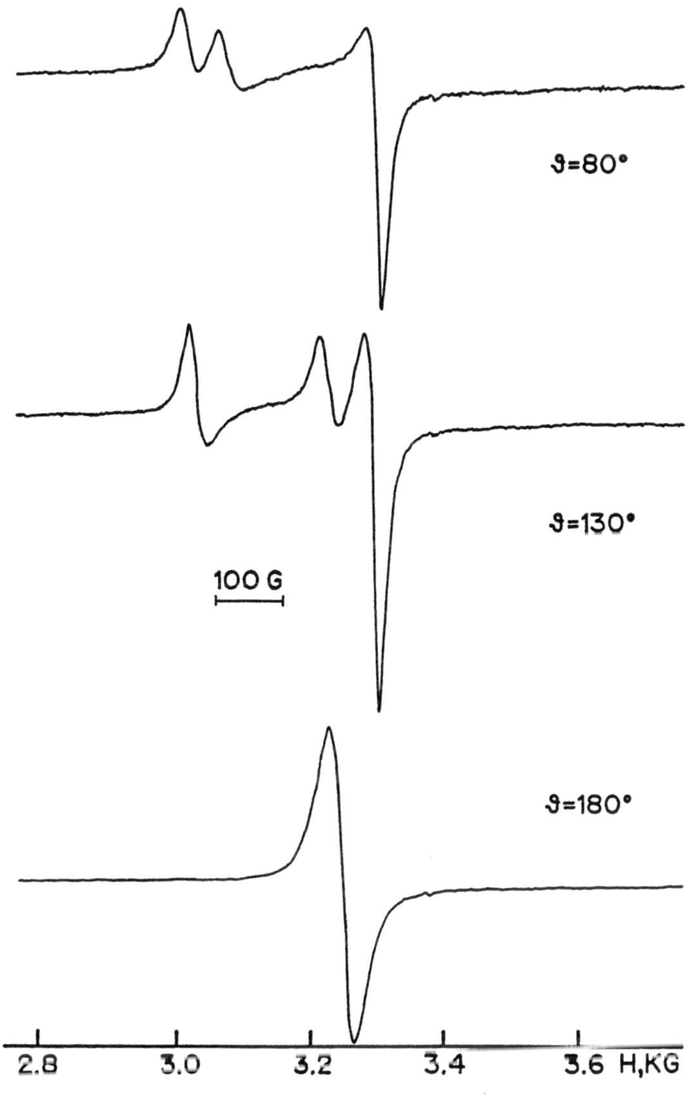

Figure 4 - ESR spectra of a LD film (30 layers) of Cu/Ba stearate after treatment with ammonia vapours. The angle θ refers to the magnetic field direction with respect to the normal to the film plane.

This agrees with the presence of $Cu[CH_3(CH_2)_{16}COO]_2 \cdot 2NH_3$, with the NH_3 molecules in the axial positions. Almost the same occurs when Cu/Ba stearate films are exposed to ammonia vapours.

Figure 4 reports the spectra of the LB film for three different orientation of the perpendicular to the plate plane (N) with respect to the magnetic field direction, as defined by the angle θ. The hyperfine structure is not resolved because of the strong spin-spin interactions (namely, Heisenberg spin-exchange among neighbour Cu^{2+} ions). Similar spectra have been reported by Messier and Marc [36] and these authors have analyzed the orientation dependence of the ESR lineshape concluding that the angle between the Cu-containing square plane and the normal to the layer is 64^0. The same conclusion is reached from our results by calculating the g factor from the relation:

$$g^2_{eff} = g_{\parallel}^2 \cos^2\alpha + g_{\perp}^2 \mathrm{sen}^2\alpha \qquad [2]$$

where α is the angle between the z molecular axis and the magnetic field direction. From the g^2_{eff} values measured as a function of the rotation, $g_{\parallel} = 2.276$ and $g_{\perp} = 2.070$, with a tilting angle of the hydrocarbon chain with respect to the normal to the film of 60 ± 5^0, are obtained. The ESR results and the data from other sources [24] allow us to suggest the structural model shown in figure 5 for the Cu/Ba stearate LB film after treatment with NH_3. This model closely resembles the one proposed for LB films of $PbSt_2$ [26], where the tilting angle is $\sim 30^0$. An appreciable difference exists, on the contrary, with the structure proposed for $MnSt_2$ LB film, which is a sheet model [6]. The presence in the film of Ba^{2+} ions should not alter the above model. The finding of ESR results similar to those of Messier and Marc [36] indicates indeed that the main effect of the Ba^{2+} ions is to increase the monolayer condensation with resultant higher stability and homogeneity of the LB film without appreciably changing its structure. The presence of an additional metal ion with a condensing effect, such as Ba^{2+}, is not necessary when LB films of manganese alcanoates are prepared. In these cases, the starting monolayers are condensed enough that their transference and superposition on a solid substrate are homogeneous and reproducible. This has been verified through isotherm determination, and with ellipsometric and ESR data [25].

Almost the same results as with $CuSt_2$ have been obtained when the copper ions are bounded to fatty acids with a hydrophobic chain longer than the C_{18} chain of the stearic acid, such as arachidic acid (C_{20}), and behenic acid (C_{22}) [25]. This allows to rule out the chain length as the determining factor in the condensation of the metal alcanoate monolayer.

In conclusion the combined use of different experimental techniques on mono- and multilayers of copper stearate clearly indicates that the dimeric Cu-Cu unit formed at the air water interface is the kinetic determining factor for the scarce stability of the transferred film.

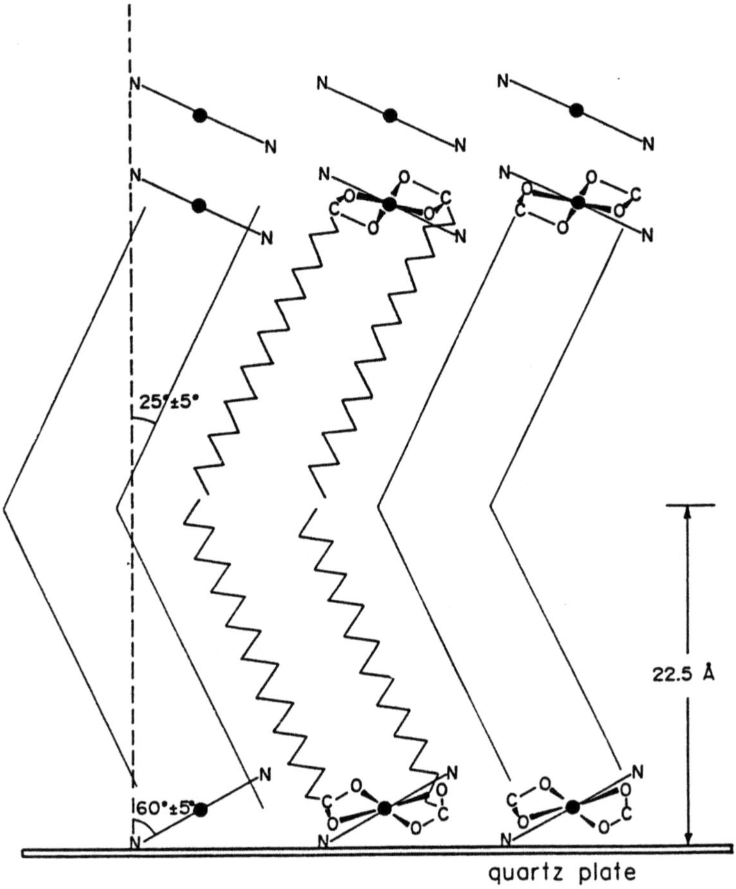

Figure 5 - Sketch of the structure proposed for copper stearate in LB film after treatment with ammonia. Ba stearate molecules are not shown since they should have a condensing effect without altering the building model.

LB FILMS OF NITROXIDE DERIVATIVES OF STEARIC ACID.

The oxazolidinyl derivatives of the stearic acid (doxyl stearic acid, DXSA):

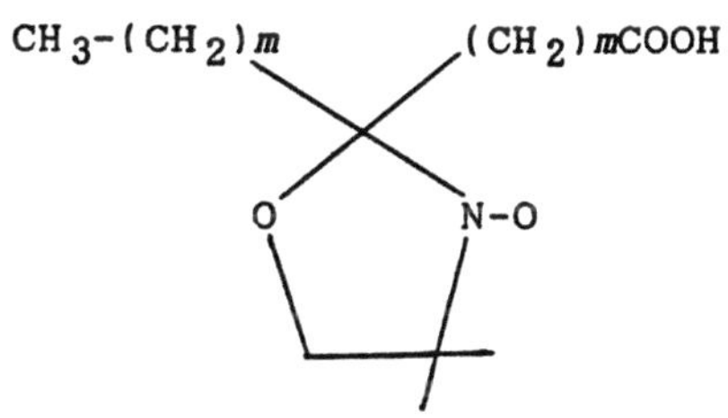

m = 3	n = 12	5-DXSA
5	10	7-DXSA
7	8	9-DXSA
10	5	12-DXSA
14	1	16-DXSA

are widely known in biological sciences because of their very large use as paramagnetic probes in the study of several properties of natural and synthetic membranes [40,41]. They give well-characterized spreading monolayers on polar subphases [22,23,42-44]. Table I reports the collapse pressures, the limiting areas, and the compressibility moduli for the most condensed phases, that are calculated from the 293 K isotherms of each nitroxide.

TABLE I

Collapse Pressures, Limiting Areas, and Compressibility Moduli of Doxvl-nitroxyde Monolayers.[a)]

System	π_{coll} (dyn/cm)	A_0 (Å^2/mol)	C_s^{-1} (dyn/cm)	$\Delta\pi$ (dyn/cm)
5-DXSA	33.9	76	42-23	12.1-32.8
7-DXSA	30.8	63	29-21	22.3-29.8
9-DXSA	26.8	36	35-16	20.3-26.1
12-DXSA	24.5	33	29-17	16.4-24.6
16-DXSA	22.1	23	66-60	>15
Stearic acid	28.4	24	290-285	23.5-28.4

a) From ref. [22].

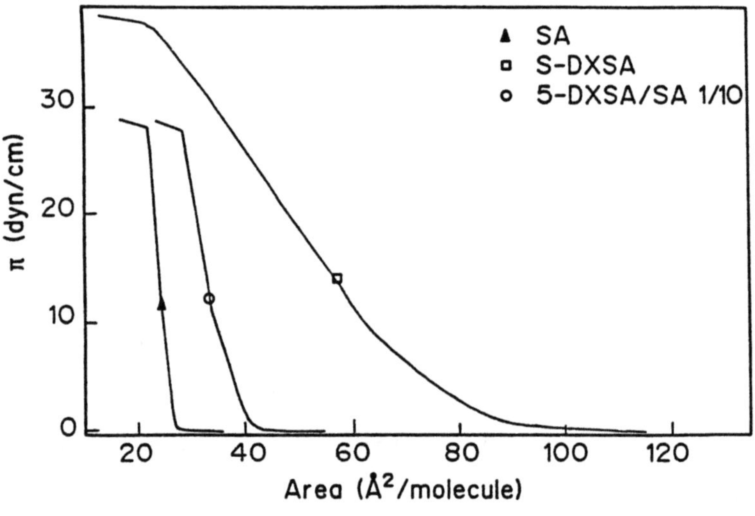

Figure 6 - Spreading isotherms at 293 K of pure 5-DXSA and of pure stearic acid and of their 1/10 binary mixture on a water subphase at pH 5.6.

From the isotherm shapes and from the data reported in the table I it results that the presence of the doxyl ring introduces large modifications in the molecular conformation at the interface and largely decreases the strength of the hydrophobic interactions among the hydrocarbon chains with a consequent film expansion. When the -COOH and the doxyl groups are sufficiently apart, as it happens in 12- and in 16-DXSA, a typical bipolar behaviour is shown at the interface. At low surface pressure, these nitroxides assume a bent structure, which turns to the usual erect conformation only at high pressure, with a strong decrease of the A_0 values, which closely resemble that of stearic acid [22,44].

The isotherms of the binary mixtures of DXSAs are intermediate between those of the pure components, with a closer similarity to that of the excess component [23]. Moreover, when a doxyl nitroxide is diluted with the diamagnetic stearic acid (SA) to obtain DXSA/SA mixtures with 1/10 ratio, the corresponding isotherm resembles the SA isotherm with only a slight expansion of the film [23], as shown, for instance, in the figure 6 for the 5-DXSA/SA 1/10 mixture.

Anyway, the A_0 values and the compressibility moduli of monolayers of pure, mixed and diluted nitroxides result , in general, determined by the size of the polar group at the interface and by the steric hindrance due to the presence of the doxyl ring. This has a relevant effect on the formation of the LB films, which may be only formed at the highest surface pressure before the monolayer collapse takes place.

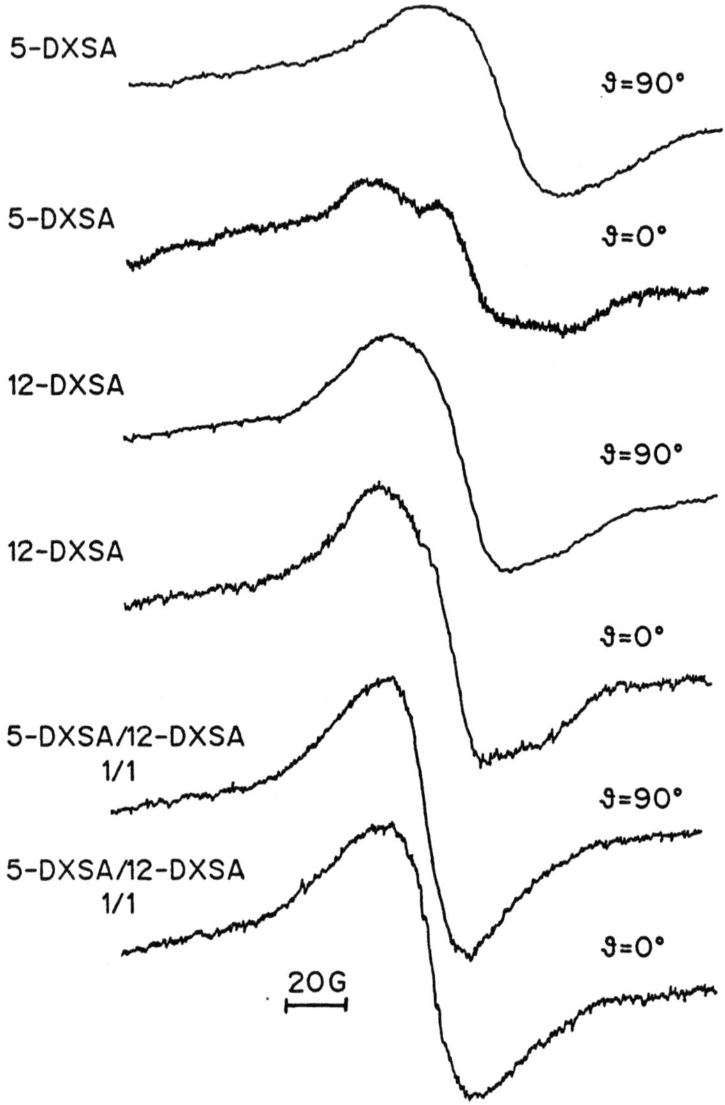

Figure 7 - ESR spectra of LB films of pure 5- and 12-DXSA nitroxides and of a their 1/1 mixture at $\theta = 0^0$ and 90^0.

In order to improve the monolayer deposition on quartz plates, the plates were preliminary covered with a stearic acid monolayer to increase the substrate hydrophobicity [22,23]. It is thus possible to build up LB films of 20 superimposed layers.

The ESR spectra of all the LB films of nitroxides we have studied (except those of LB films of nitroxides diluted with SA) are dominated by strong spin-spin interactions (dipole-dipole and Heisenberg spin-exchange) as it is exemplified in figure 7. The spectra are broad and the linewidth depends on the position of the five-membered ring and on the orientation with respect to the magnetic field defined by the angle θ (see above; in this case the normal to the plate is assumed to coincide with the z molecular axis of the nitroxide, i.e. with the hydrocarbon chain direction).

The ESR spectra of the LB films prepared by transference of the monolayers of DXSA diluted in SA closely resemble those of nitroxides slightly concentrated in very viscous solvents, i.e. in very slow motion conditions, with only a very limited averaging of the magnetic anisotropies [45]. The LB film of 1/10 5-DXSA/SA has also been analyzed through pulse wave ESR [46]. The study of the ESE signals has been carried out on the basis of the theory of Mims [47] and this has allowed the evaluation of the distance among the unpaired electron and the proton nuclei of both doxyl ring and aliphatic chain. The agreement between the calculated values and those expected from a structural model is quite good.

A computation of the c.w. ESR spectra of nitroxide LB films, based on the procedure reported by Freed [45] which takes into account the modulation of the magnetic anisotropies g and A, through a correlation time for the motion τ, has been used to obtain both molecular mobility and spin-spin interactions. The linewidth to be simulated is given by [48]:

$$\Delta B = \alpha' + \alpha'' + \beta m_I + \gamma m_I^2 \qquad [3]$$

where α', β, and γ depend on the magnetic anisotropies and on the spectral densities [45,48], whereas α'' contains non-motional terms, including dipole-dipole broadening, ΔB_{dip}, and exchange broadening, ΔB_{exc}, in the slow exchange conditions. The exchange narrowing effects in the fast exchange limits are accounted by simply summing a lorentzian single line with a width related to the spin-exchange frequency predicted by the modified Bloch equations [49], in the sense that the higher is the linewidth the lower is the exchange frequency. The accuracy of the fits is shown by the example reported in figure 8. Table II reports representative data of the best fit parameters for the simulation of the ESR spectra of LB films of pure nitroxides and of their mixtures with SA. These results may be compared with those obtained with bulk nitroxides and with disordered multilayered materials obtained when the collapse pressure of the monolayer is reached [22].

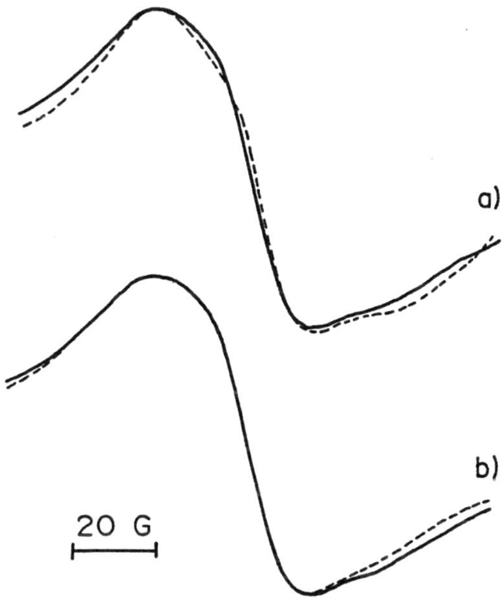

Figure 8 - Experimental (full lines) and computed ESR spectra (dashed lines) of LB films of 12-DXSA (θ = 90^0): a) simulation according to Freed procedure neglecting Heisenberg spin exchange (τ = 3x10^{-7}s; $T_{2,0}$ = 23 G); b) simulation as a sum of spectrum a) with a lorentzian single exchange-narrowed line withΔB = 35 G.

TABLE I·I

Best-fit Parameters for the Simulation of the 298 K ESR Spectra of LB Films of Single DXSA and of DXSA Diluted in Stearic Acid.[a)]

LB Film	τ,sx10^7 θ=90^0	θ=0^0	Dipole-dipole broadening anisotropy	ΔB_{exc}, G
5-DXSA	3	1	15	50
7-DXSA	3	1	8	60
12-DXSA	1	0.3	3	35
16-DXSA	0.3	0.03	2	35
5-DXSA/SA (1/10)	1	0.2	-	8
16-DXSA/SA (1/10)	0.5	0.1	-	4

a) From refs. [22,23].

In summary, the ESR spectra of the LB films built up with pure nitroxides are dominated by strong spin-spin interactions. In particular, the Heisenberg spin-exchange is always isotropic and the exchange frequencies result lower than in bulk and in collapsed systems. This suggests a more open structure in the film with higher mean distances among the unpaired electrons. The exchange frequencies in the films increase with the distance between -NO and -COOH groups, thus indicating a mobility gradient along each layer. The dipole-dipole interactions are anisotropic when pure nitroxide films are analyzed, which accounts for a shape anisotropy. The observed decrease of the anisotropy with the -COOH/NO distance agrees with the increase of the mobility along the chain suggested by the exchange frequencies. The dipole-dipole interactions are practically absent in the diluted DXSA films, where only weak and isotropic Heisenberg spin-exchange interactions are observed. In both kinds of LB systems, the wobbling frequencies, as valued from the correlation times for the motion, are in the range 10^7-$3x10^8$ s^{-1}, as expected for multilayer systems and of the same order of magnitude as those reported for natural and synthetic membranes.

EXPERIMENTAL SECTION

Experimental details are given only for the copper stearate systems. For nitroxide systems refer to [22,23].

Materials. Stearic acid (Erba, purity >99%) solutions in chloroform (Erba, purity >99.5%) were spread on twice-distilled water (resistivity > 18 MΩ.cm) subphases for the monolayer and for the LB formation. The aqueous subphases were as it follows: a) 3.3×10^{-3} mol/L solution of $Cu(NO_3)_2$ (Erba) with pH adjusted to 5.6 with NaOH; b) 3×10^{-5} $BaCl_2$ (Merck) and 2×10^{-6} mol/L $Cu(NO_3)_2$ with pH adjusted to 7.2 with $KHCO_3$ (Merck).

Copper stearate was prepared and purified, according to the procedure given in ref. [50].

Techniques. The surface pressure/area isotherms were obtained using the Langmuir method with a computer controlled Lauda balance. LB films were prepared with the Langmuir Trough 4 (Joyce-Loebl). The dimensions of the quartz plates were 40x4x0.5 mm. The transfer rate on the plate was 3 mm/min.

Ellipsometric measurements were made with a Rudolf Research ellipsometer model 437-02. Quartz plates coated with a vacuum evaporated chromium layer were used.

ESR spectra were registered with a Bruker ESR spectrometer model 200D operating in the X-band and equipped with an Aspect 2000 data handling system for signal accumulation and magnetic parameter valuation. The temperature was varied by using the Bruker ST 100/700 variable temperature accessory.

Acknowledgments.

Thanks are due to the Italian Council of Research (CNR) and to the Italian Ministero della Universita' (MURST) for the financial support.

REFERENCES

1) Roberts, G.G.; *Adv. Phys.*, **1985**, *34*, 475.

2) Swalen, J.D.; Allara, D.L.; Andrade, J.D.; Chandross, E.A.; Garoff, S.; Israelashvili, J.; McCarthy, T.J.; Murray, R.; Pease, R.F.; Rabolt, J.F.; Wynne, K.J.; Yu, H. *Langmuir*, **1987**, *3*, 932.

3) Möbius, D. *Acc. Chem. Res.* **1981**, *14*, 63.

4) Pomerantz, M. *Solid State Comm.* **1978**, *27*, 1413.

5) Pomerantz, M.; Dacol, F.H.; Segmüller, A. *Phys. Rev. Lett.* **1978**, *40*, 246.

6) Pomerantz, M.; Segmüller, A. *Thin Solid Films*, **1980**, *68*, 33.

7) Ruaudel-Teixier, A.; Barraud, A.; Vandevyver, M.; Belbeoch, B.; Rouliay, M. *J. Chim. Phys. (France)*, **1985**, *82*, 711.

8) Barraud, A.; Ruaudel-Teixier, A.; Vandevyver, M.; Lesieur, P. *Nouveau J. Chim.* **1985**, *9*, 365.

9) Richard, J.; Vandevyver, M.; Lesieur, P.; Ruaudel-Teixier, A.; Barraud , A.; Bozio, R.; Pecile, C. *J. Chem. Phys.* **1987**, *86*, 2428.

10) Vandevyver, M.; Barraud, A. *J. Molec. Electron.* **1988**, *4*, 207.

11) Ikegami, K.; Kuroda, S.; Saito, K.; Sugi, M.; Nakamura, T.; Matsumoto, M.; Kawabata, Y.; *Phys. Rev. B*, **1987**, *35*, 3667.

12) Ikegami, K.; Kuroda, S.; Sugi, M.; Iizima, S.; Nakamura, T.; Matsumoto, M.; Kawabata, W.; Saito, G. *Synth. Meth.* **1987**, *19*, 699.

13) Iwasaki, T.; Wakabayashi, H.; Ishii, T.; Iriyama, K. *Appl. Phys. Lett.* **1984**, *45*, 1089.

14) Kuroda, S.-I.; Sugi, M.; Iizima, S. *Thin Solid Films*, **1985**, *133*, 189.

15) Pace, M.D.; Berger, W.R.; Snow, A.W. *J. Magn. Reson.* **1987**, *75*, 73.

16) Cook, M.J.; Dunn, A.J.; Gold, A.A.; Thomson, A.J.; Daniel, M.F. *J. Chem. Soc. Dalton Trans.* **1988**, 1583.

17) Cook, M.J.; Daniel, M.F.; Dunn, A.J.; Gold, A.A.; Thomson, A.J. *J. Chem. Soc. Chem Commun.* **1986**, 863.

18) Palacin, S.; Ruaudel-Teixier, A.; Barraud, A. *J. Phys. Chem.* **1986**, *90*, 6237.

19) McArdle, C.B.; Ruaudel-Teixier, A. *Thin Solid Films*, 1985, *133*, 93.

20) Quinn, P.J. *The Molecular Biology of Cell Membranes*; McMillan: London, 1976.

21) Kinnunen, P.K.J.; Virtanen, J.A.; Tulkki, A.P.; Ahuja, R.C.; Möbius, D. *Thin Solid Films*, 1985, *132*, 193.

22) Bonosi, F.; Gabrielli, G.; Martini, G.; Ottaviani, M.F. *Langmuir*, 1989, *5*, 1037.

23) Martini, G.; Bonosi, F.; Ottaviani, M.F.; Gabrielli, G. *Thin Solid Films*, 1989, *178*, 271.

24) Gabrielli, G.; Puggelli, M.; Ferroni, E.; Carubia, G.; Pedocchi, L. *Colloids Surfaces*, 1989, *41*, 1.

25) Bettarini, S.; Bonosi, M.; Gabrielli, G.; Martini, G. to be published.

26) Stephens, J.F.; Tuck-Lee, C. *J. Appl. Crystallogr.* 1969, *2*, 1.

27) Pillbrow, J.R.; Smith, T.D. *Coord. Chem. Rev.* 1974, *13*, 173.

28) Sharrock, P.; Melnik, M. *Can. J. Chem.* 1985, *63*, 52.

29) Yablokov, Yu.V.; Ablov, A.V. *Dokl. Akad. Nauk SSSR (English transl.)* 1962, *144*, 375.

30) Sharrock, P.; Dartiguenave, M.; Dartiguenave, Y. *Bioinorg. Chem.* 1978, *9*, 3.

31) Bleaney, B.; Bowers, K.D. *Proc. Roy. Soc. Ser. A.*, 1952, *214*, 451.

32) Harish, S.P.; Sobhanadri, J. *Inorg. Chim. Acta*, 1985, *108*, 147.

33) Lewis, J.; Mabbs, F.E.; Roylston, L.K.; Smail, W.R. *J. Chem. Soc. A*, **1969**, 291.

34) Kakuta, N.; Kasuzaka, A.; Yamazaki, A.; Miyahara, K. *J. chem. Soc. Faraday Trans, 1*, **1984**, *80*, 3245.

35) Chollet, P.A. J. Phys. C., *Solid State Phys.* **1974**, *7*, 4127.

36) Messier, J.; Marc, G. *J. Physique*, **1971**, *32*, 799.

37) Ablov, A.V.; Yablokov, Yu.V.; Zheru, I.I. *Dokl. Akad. Nauk SSSR (English transl.)*, **1961**, *141*, 1116.

38) Yablokov, Yu.V.; Gavrilov, V.V.; Milkova, L.N.; Ablov, A.V. *J. Struct. Chem.* **1971**, *12*, 216.

39) Martin, R.L. *J. Chem. Soc. A*, **1958**, 1394.

40) Smith, J.C.P.; Butler, K.W. In *Spin Labeling. Theory and Applications;* Berliner, L.J. Ed.; Academic: New York; Vol. 1, pp. 411-451. Griffith, O.H.; Jost, P.C. *Ibid.* pp. 453-523.

41) Marsh, D. In *Membrane Spectroscopy;* Grell, E., Ed.; Springer: Berlin, 1981; pp. 51-142.

42) Cadenhead, D.A.; Müller-Landau, F. *Adv. Chem. Ser.* **1975**, *144*, 294.

43) Cadenhead, D.A.; Müller-Landau, F. *J. Colloid Interface Sci.* **1973**, *49*, 131.

44) Cadenhead, D.A.; Müller-Landau, F. *Biochim. Biophys. Acta*, **1973**, *307*, 279.

45) Freed, J.H. In Spin Labeling. *Theory and Applications;* Berliner, L.J., Ed.; Academic: New York, 1976; Vol. 1, pp. 53-132.

46) Romanelli, M.; Bonosi, F.; Chen, X.; Kevan, L. Langmuir, submitted.

47) Mims, W.B. *Phys. Rev. B*, **1972**, *5*, 2409.

48) Kivelson, D. *J. Chem. Phys.* **1960**, *33*, 1094.

49) Sackman, E.; Traüble, H. *J. Amer. Chem. Soc.* **1972**, *94*, 4492.

50) Dobry, A. *J. Chem Phys.* **1954**, *58*, 576.

Comparison of the Superoxide Scavenging Activity of Various Manganese Compounds by Use of ESR Spectroscopy

S. RIVOMANANA°, M. MASSOL°, P. DERACHE°° and F. NEPVEU°°°

Université Paul Sabatier: °Laboratoire de Chimie Bioinorganique - INSERM U-305, 38, rue des 36 Ponts, F-31400 Toulouse; °° Faculté des Sciences Pharmaceutiques, Laboratoire de Biochimie, 35, Chemin des Maraîchers, F-31052 Toulouse Cedex; °°°Laboratoire de Chimie Analytique, 31, Allées Jules Guesde, F-31400 Toulouse.

INTRODUCTION

Aerobic organisms synthesize enzymes devoted to the prevention and repair of oxidative damage. Superoxide dismutases (SOD) are able to disproportionate superoxide radicals and the presence of metal ions at their actives sites (Cu/Zn - SOD, Mn-SOD) led to a search for small molecular weight complexes with similar activities (1 - 4). Concurrently, manganese ions are used by *Lactobacillus plantarum* to dispose of the superoxide anion but this organism has none of the known superoxide dismutases. In such bacteria, manganese is associated to undetermined organic molecules to scavenge the superoxide anion (5 - 6). Based on such observations, we investigated the possible SOD-like activities of various manganese complexes.

In this paper, we report on the superoxide scavenging properties of manganese(II) ion complexed with lactic, malonic, mandelic or anthranilic acids. The SOD-like activity of these compounds have been carried out with Electron Spin Resonance spectroscopy (ESR) and the technique of spin trapping. DMPO (5,5-dimethyl-1-pyrroline-N-oxide) and POBN (α-(4-pyridyl-1-oxide)-N-tert-butylnitrone) have been used as spin traps. The results are discussed in relation to the nature of the spin trap and of the complexing organic acid. Partial results with DMPO have been previously reported (7).

MATERIAL AND METHODS

Chemical compounds were obtained from commercial sources and used without further purification. DMPO, POBN and enzymes were obtained from Sigma Chemical Co. The DMPO solution was purified with activated charcoal and checked for absence of any ESR contaminating signal. The following experimental conditions have been taken to minimize artifacts or secondary reactions: the acetaldehyde / Xanthine oxidase (Ach/XO) reaction has been used as superoxide generating system; diethylenetriaminepentacetic acid (DTPA) was included in the phosphate buffer to remove metal ion impurities like iron and prevent hydroxyl radical formation and DMPO oxidation (8 - 10).

Final concentrations were: with DMPO: 10 mM Ach, 12 mU/ml XO, 15 μM DTPA, 50 mM DMPO, final volume 2 ml with phosphate buffer, KH_2PO_4 0.1 M, pH = 7.4; with POBN: 100 mM Ach, 60 mU/ml XO, 15 μM DTPA, 50 mM POBN final volume 2 ml with phosphate buffer, KH_2PO_4 0.1 M, pH = 7.4. Reactions were initiated by addition of Ach (time = 0 min.). After different development times

B. Catoire (Editor)
Electron Spin Resonance (ESR) Applications
in Organic and Bioorganic Materials

(DMPO: 5 min; POBN: 5, 15, 30 and 45 min) ESR spectra were recorded at room temperature using a BRUCKER ER200D X-band spectrometer. The instrument settings were: microwave frequency 9.72 GHz, central field set 3500 G, field scan 200 G, modulation of amplitude 1 Gpp, modulation frequency 100 kHz, time constant 0.5 s, cavity TM 110.

The manganese compounds have been prepared from manganese (II) chloride or manganese(II) acetate and the following carboxylic acids: lactic (LAC), malonic acid (MAL), mandelic acid (MAN), anthranilic acid (ANT). The microanalytical data and physico-chemical results are consistent with the following formula:

lactic acid: $(Mn(C_3H_5O_3)_2(H_2O)_2)$:(Mn-LAC):**1**
malonic acid: $(Mn(C_3H_2O_4)(H_2O)_2)$:(Mn-MAL):**2**
mandelic acid: $(Mn(C_8H_7O_3)_2)$:(Mn-MAN):**3**
anthranilic acid: $(Mn(C_7H_6NO_2)_2)$:(Mn-ANT):**4**

Formula for compounds **1** and **2** are in agreement with X-ray structural data (11 - 12).

RESULTS AND DISCUSSION

Figure 1(A) illustrates ESR spectrum of DMPO-OOH spin adduct recorded 5 min after adding Ach to the reaction mixture of a blank solution. When incubation time is increased (7 to 10 min) ESR spectrum can exhibit surperimposition of signals of both superoxide adduct DMPO-OOH and hydroxyl adduct DMPO-OH. This demontrates the unstable nature of DMPO-OOH as it has been reported (13 - 15). In order to confirm the formation of DMPO-OOH signal from superoxide radical and to verify that generation of DMPO-OH was due to a decomposition of DMPO-OOH when incubation time was greater than 5 min, 0.75 U/ml SOD was added to the blank solution. Formation of both spin adducts was decreased with addition of excess of SOD, preventing DMPO-OOH and DMPO-OH signal formation. Use of acetaldehyde as substrate (and not xanthine or hypoxanthine) and other experimental conditions were necessary to keep the intensity of DMPO-OOH signal and avoid the spontaneous decomposition of DMPO-OOH into DMPO-OH within the 5 min incubating time before recording.

In following experiments, each tested manganese complex was added at various concentrations (5 to 50 µM). The DMPO-OOH signal gradually disappeared as the concentration of complex was increased. From these tests, the concentration giving an 50% signal intensity inhibition, in respect to the blank ESR signal intensity, was determined.

Fig 1 (A, B, C) illustrates a typical assay with compound **4**. Manganese complexes could generate the hydroxyl radical (OH°) by catalyzing the decomposition of H_2O_2 formed during the superoxide dismutation reaction. However, in our experimental conditions, (Fig. 1 (B, C)), DMPO-OH was never observed.

Table 1 compares the manganese complex concentration for obtaining an 50% signal intensity inhibition. The compounds **1**, **2**, **4**, exhibit a significative difference in their superoxide scavenging properties.

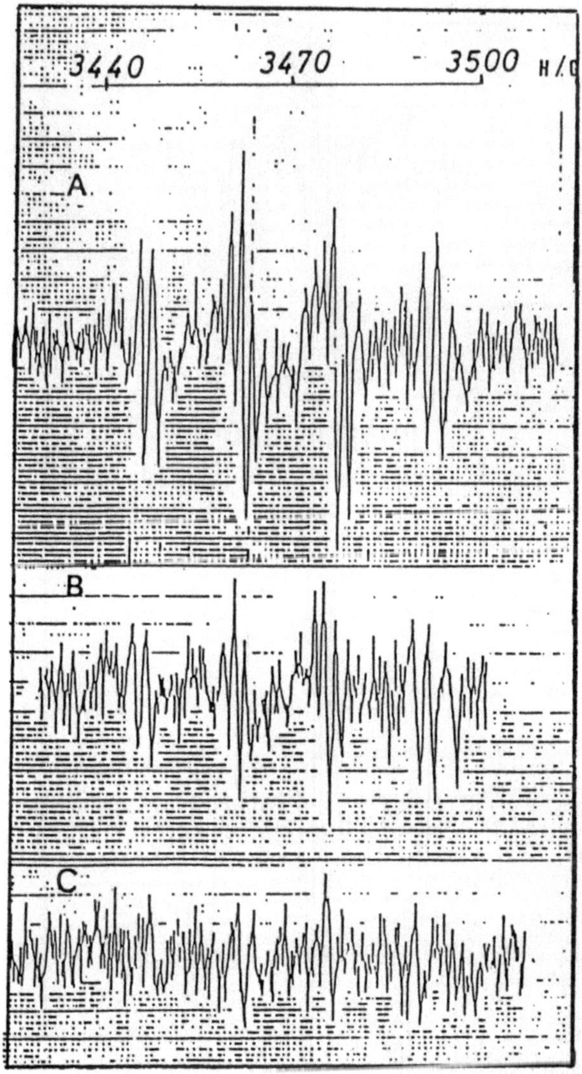

Fig.1 (A): ESR spectrum obtained 5 min after adding Ach (10mM) to a blank solution containing 12 mU/ml XO, 5 µM DTPA, 50 µM DMPO, phosphate buffer pH 7.4 to 2 ml, final volume.
(B): conditions same as in (A) except that compound 4 (20 µM) was added;
(C): conditions same as in (A) except that compound 4 (50 µM) was added.
DMPO-OOH: g = 2.0049, a_N = 16.41 G, a_H^{α} = 14.62 G, a_H^{γ} = 2.15 G.

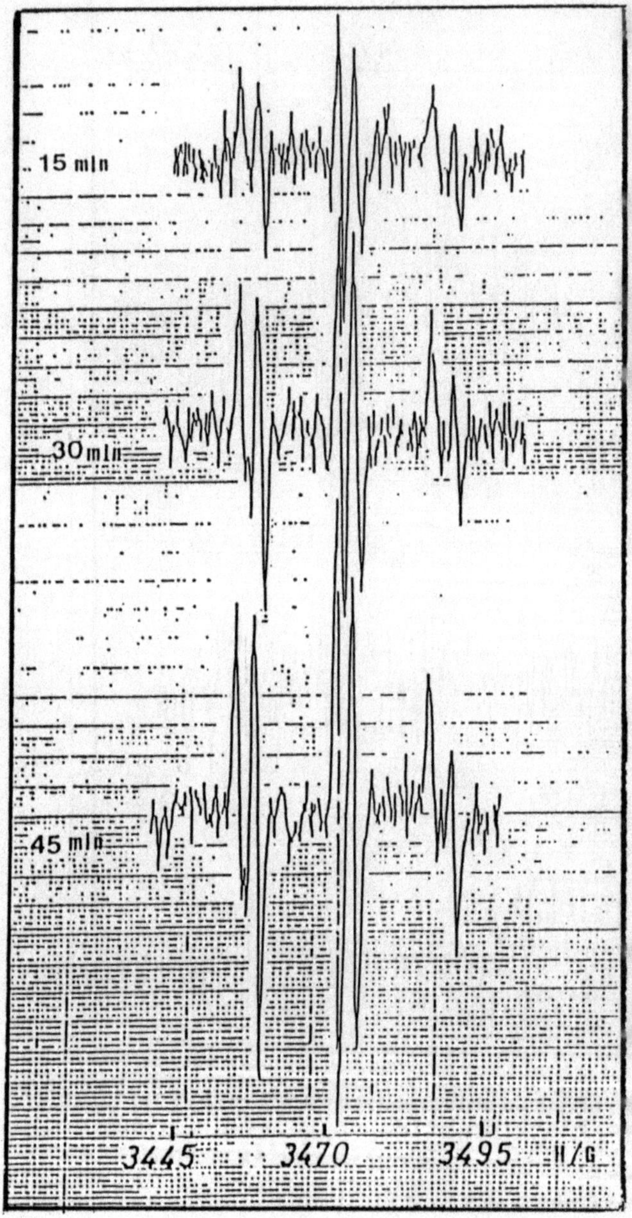

Fig.2: ESR spectra obtained 15, 30 and 45 min after adding Ach (100mM) to a blank solution containing 60 mU/ml XO, 15 µM DTPA, 50 mM POBN, phosphate buffer pH = 7.4 to 2 ml, final volume.
POBN-OOH: g = 2.0052, a_N = 15.68 G, a_H^{α} = 3.23 G

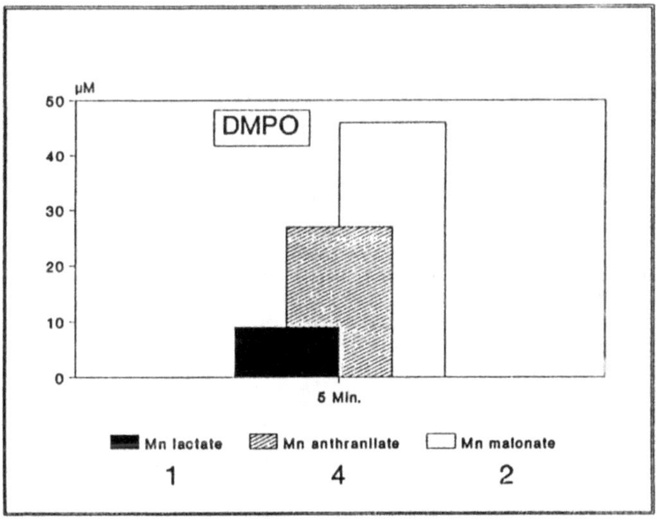

Table 1: Comparison of the manganese complex concentration for an 50% signal intensity inhibition for compounds **1**, **2** and **4** with DMPO as spin trap (incubation time: 5 min)

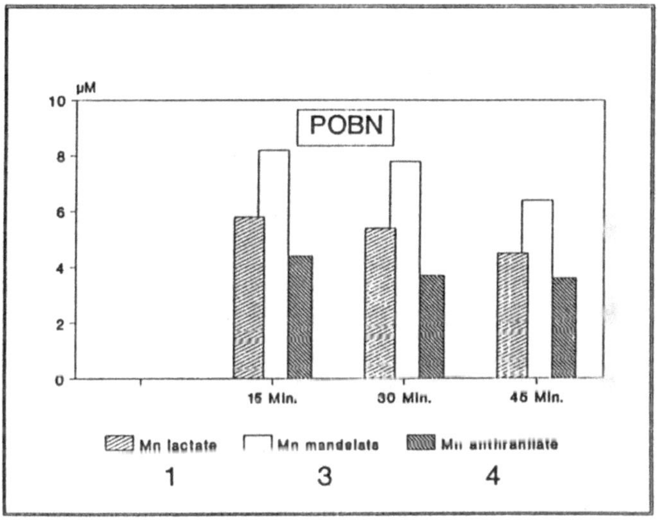

Table 2: Comparison of the manganese complex concentration for an 50% signal intensity inhibition for compounds **1**, **3** and **4** with POBN as spin trap (incubation time: 15, 30 and 45 min).

Fig.2 shows ESR spectra of POBN-OOH spin adduct recorded at 15, 30 and 45 min after adding Ach to the blank reaction mixture. This figure evidences the greater shelf live of POBN spin adducts. However, under our experimental conditions, it is necessary to give a development time of 15 min to get nearly the same signal intensity with POBN as with DMPO after 5 min and with higher concentrations in enzyme and substrate. These results illustrate the lesser reactivity of POBN with superoxide anion under our experimental conditions compared to DMPO as it has been reported (8). To confirm the formation of superoxide radical, 0.75 U/ml SOD was added to the solution, the formation of POBN spin adducts was prevented. Since the typical six lines signal of POBN-OOH is not easily distinguishable from the six lines ESR signal of POBN-OH, it was not possible to observe the eventual decomposition of POBN-OOH into POBN-OH. In the following experiments, each tested manganese complex was added at various concentrations (0.5 to 20 µM) to the reaction mixture. ESR spectra were recorded after reaction development times of 15, 30 and 45 min. This means a new reaction mixture solution by incubation time, by tested metallic compound concentration, and by manganese complex.

Table 2 compares the manganese complex concentration necessary to get an 50% signal intensity inhibition. In the reported experimental conditions and taking into account experimental errors, the tested compounds do not exhibit significative difference in their superoxide scavenging properties when POBN is used as spin trap.

CONCLUSION

The lesser reactivity of POBN with superoxide anion favours the superoxide scavenging reaction and explains that (i) the manganese complex concentration to get an 50% signal intensity is very low, (ii) all tested compounds appear to have the same activity (Table 2).

The greater reactivity of superoxide anion with DMPO leads to a competition between the spin trapping reaction and the superoxide scavenging reaction and allows a differentiation in the reactivity of metallic compounds (Table 1).

These results demonstrate the superoxide scavenging properties of tested manganese compounds, the most reactive being compounds **1** or **4**. These compounds do not generate the hydroxyl radical OH° in the reaction mixture.

Thus, in conclusion, DMPO appears to be an useful spin trap for the comparison of SOD-like properties of metallic compounds taking into accounts its disadvantages: the experiments have to be run in specific experimental conditions, in a very short time, and with sufficient controls to validate the results.

BIBLIOGRAPHIE

1. SORENSON J.R.J., *in* Biology of Copper Complexes, eds J.R.J. Sorenson, Clifton, New Jersey, Humana Press, 1987, pp. 3-16.

2. DEUSCHLE, U., and WESER, U., Inorg. Chim. Acta, 1984, **91**, 237-242.

3. HUBER, K.R., SRIDHAR, R., GRIFFITH, E.H., AMMA, E.L., and ROBERTS, J., Biochim. Biophys. Acta, 1987, **915**, 267-276.

4. RABINOWITCH, H.D., ROSEN, GM., and FRIDOVICH, I., J. Free Radic. Biol. Med., 1989, **6**, 45-48.

5. ARCHIBALD, F.S., and FRIDOVICH, I., J. Bacteriol., 1981, **146**, 928-936.

6. ARCHIBALD, F.S. and FRIDOVICH, I., Arch. Biochem. Biophys., 1982, **214**(2), 452-463.

7. NEPVEU, F., DUCROCQ, A., and DERACHE, P., *in* Trace Element Analytical Chemistry in Medecine and Biology, eds P. Brätter and P. Schramel. Berlin: Walter de Gruyter, 1988, vol.5, 601-606.

8. FINKELSTEIN, E., ROSEN, G.M., and RAUCKMAN, E.J., Arch. Biochem. Biophys., 1980, **200**, 1-16.

9. FLOYD, R.A., Biochim. Biophys. Acta, 1983, 756, 204-216.

10. SAMUNI, A., KRISHNA, M., RIESZ, P., FINKELSTEIN, E., and RUSSO, A., J. Free Radic. Biol. Med., 1989, **6**, 141-148.

11. LIS, T., Acta Cryst., 1982, B**38**, 937-939.

12. LIS, T., and MATUSZWEWSKI, J., Acta Cryst., 1979, **B35**, 2212-2214.

13. FINKELSTEIN, E., ROSEN, G.M., and RAUCKMAN, E.J., Mol. Pharmacol., 1979, **16**, 676-685.

14. FINKELSTEIN, E., ROSEN, G.M., and RAUCKMAN, E.J., Mol. Pharmacol., 1982, **21**, 262-265.

15. BUETTNER, G.R., and OBERLEY, L.W. Biochem. Biophys. Res. Comm., 1978, **83**(1), 69-74.

Thermal Decomposition Kinetics of Polypropylene Peroxides in the Polymer Matrix by ESR

B. CATOIRE*, V. VERNEY (a), A. MICHEL (a)

* INSTITUT TEXTILE DE FRANCE - Section Lyon
BP 60 - 69132 ECULLY CEDEX, FRANCE

(a) LABORATOIRE DES MATERIAUX ORGANIQUES - C N R S
BP 34 - 69390 VERNAISON, FRANCE

ABSTRACT

Ozonization creates peroxidic species, peroxides and hydroperoxides in the bulk of polypropylene powder. Their characterization and chemical determination is delicate (135°C in decalin). Moreover, the thermal decomposition kinetics of the reactive groups at various temperature cannot be easily followed by usual methods. In order to follow the thermally induced radicals in the polymer matrix, we used the "spin trapping technique" in E S R. Concurrently we looked at the peroxidic decomposition exothermicity using a differential mode technique in D S C. The E S R data show that the spin trap technique can be used successfully to follow the thermal decomposition kinetics. The activation energy (100 kJ/mol) and rate constant values obtained with this technique are in good agreement with those found in the literature by other techniques. The relative thermogram technique in D S C allows to get the information about peroxidic decomposition exothermicity.

INTRODUCTION

It is well known that ozonization creates peroxidic species in the bulk of polypropylene (PP). Their characterization and chemical determination have been extensively studied (1, 2, 3, 4, 5). Nevertheless, the polypropylene solubilization is a delicate operation (135°C in decalin) and makes the characterization of the species formed quite difficult. Moreover, the thermal decomposition kinetics of the reactive groups at various temperatures cannot be easily followed by usual methods.

The thermally induced radicals are generally too short-lived species and cannot be observed by electron spin resonance (E S R) unless being previously stabilized by other molecules (Spin-Trap).

In a previous work, "E S R Spin-Trapping" was used to study thermal decomposition of peroxidized polypropylene ozonized under various conditions (6). The purpose of this paper is to apply this technique to obtain kinetic parameters from thermal decomposition isotherms.

Concurrently, Differential Scanning Calorimetry (D S C) technique was carried out with control P P as a reference and ozonized P P as a sample in order to isolate peroxidic decomposition exothermicity.

B. Catoire (Editor)
Electron Spin Resonance (ESR) Applications
in Organic and Bioorganic Materials

EXPERIMENTAL

Polypropylene and ozonization

Polypropylene used in this study was kindly supplied by "Naphtachimie" ; its weight-average molecular weight is 289 000 and its number-average molecular weight 48 500. It is a powder with an average diameter of 290 µm and stabilizers-free.

Ozonization was carried out in a fluidized bed with an ozone oxygen/nitrogen gas mixture (O_3 = 6.2 mg/l) during 300 min at 70°C.

Peroxides content was determined by an iodometric method (7). Its value is 35 . 10^{-6} mol/g.

Experimental conditions

Polypropylene samples (control and ozonized) were dipped into CH_2Cl_2 solution containing 0,2 mol/l "Spin Trap" during 1 hour at 40°C. CH_2Cl_2 is a swelling agent.

Spin-Traps used are : N-tert-butylphenylnitrone (PBN) and N-tert-butyl-4-α-pyridyl-1-oxide (POBN) (Janssen Chemical).

After inclusion, samples were dried at 35°C for 3 hours to get rid of all the solvent letf in the structure and observed without washing.

Inside E S R cavity, samples including POBN and PBN were heated at 75, 83, 95 and 115°C varying the experiment durations. All spectra, in our experiments, were recorded after returning to room temperature and not under heating conditions. Some complementary runs have allowed us to be sure that nitroxide radicals induced from nitrone were rather stable in the limits of time and temperature required for the experiments.

D S C measurements were carried out at various heating rates (5, 10 and 20°C/min). Reference cell contains the same weight of control P P as ozonized P P in the sample cell. To characterize samples before and after heat treatment, heating rate was 10°C/min with an empty reference cell.

RESULTS

E S R Results

Figure 1 shows E S R spectra of ozonized P P including PBN or POBN and heated at 115°C during 15 min. These spectra were recorded after returning to room temperature. The curves consist of three wide and asymmetrical lines characteristic for strongly immobilized nitroxides in the amorphous zones of P P. Under similar conditions we did not observe any significant response with control P P.

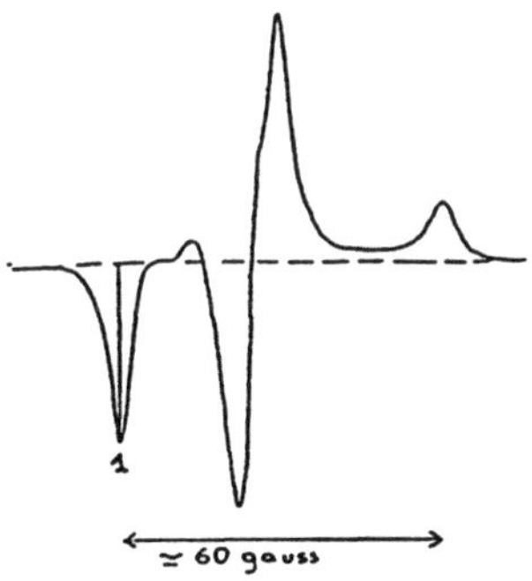

Figure 1 : E S R spectrum of P P including POBN or PBN after heating at 115°C - 15 min -
Spectrum was recorded at room temperature

We emphasize that the area located under the absorption curve is proportional to the number of spins. As long as the absorption curve has the same shape, the area is proportional to the derivate curve amplitude. The line referenced "1" on the spectra allows comparison of the intensities (other things being equal).

As long as the spectra are recorded at room temperature, we observed similar shapes of curves. This permits to get rid of temperature-dependent mobility effects of macromolecular segments in the "amorphous zones", which could after our measurements. let us remind that each isothermal determination was performed using the same sample. For a same temperature, heating was periodically stopped to get spectra recorded at room temperature. We checked that heating and cooling rates were fast enough to have cumulative temperature.

Thermal decomposotion isotherms :

Graph 1 shows that thermal decomposition isotherms of peroxidic species at 75, 83, 95 and 115°C for ozonized P P including POBN, and graph 2 for ozonized P P including PBN. Peak intensities (cf. figure 1, peak "1") are reported in ordinate (arbitrary units) and treatment durations in abscissa. Below 75°C induced radical content is too weak for significant measurements.

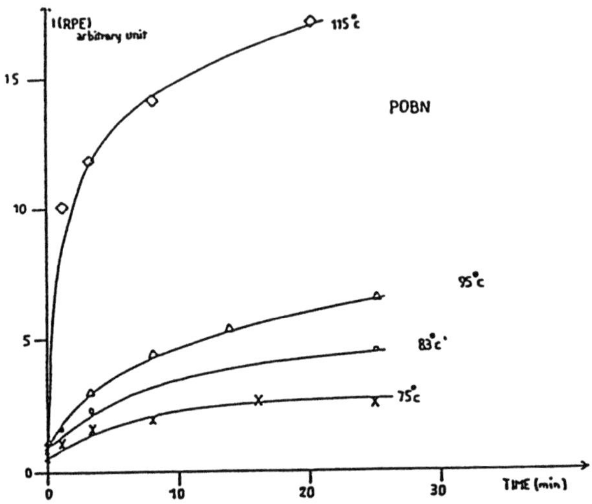

<u>Graph 1</u> : *Isotherms of the thermal decomposition of peroxidic species at 75, 83, 95 and 115 °C for ozonized P P including POBN*
E S R signal intensity (arbitrary unit) is plotted versus heating time

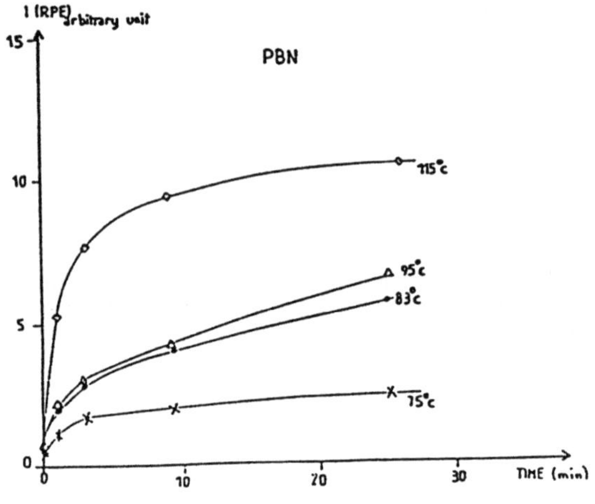

<u>Graph 2</u> : *Isotherms of the thermal decomposition of peroxidic species at 75, 83, 95 and 115 °C for ozonized P P including PBN*
E S R signal intensity (arbitrary unit) is plotted versus heating time

From these two graphs (1 and 2) we measured inital rates (V_i). Values are reported in Table 1 :

T (°C)	1/T x 10' (K^{-1})	V_i (POBN) (a.u./min)	LnV_i (POBN)	V_i (PBN) (a.u./min)	LnV (PBN)
75	2,87	0,38	- 0,97	0,38	- 0,97
83	2,81	0,66	- 0,41	1,35	- 0,30
95	2,72	2,82	+ 1,04	1,92	+ 0,65
115	2,58	13,0	+ 2,56	12,0	+ 2,48

Table 1

D S C Results

D S C measurements were carried out with the same weight of control P P in the reference cell as ozonized P P in the sample cell. We shall call these measurements "relative thermograms" in contrast with those obtanied "classically" (empty reference cell). Furthermore, we checked that the same weight (ozonized or control P P) in the two cells gives a flat recording.

Control or ozonized P P "classical thermograms" are shown on graph 3. We notice that ozonized P P melting temperature is 3°C lower than that of control P P (161°C instead of 164°C). moreover, there are no thermal transitions before the melting endotherm.
On graph 4 are drawn "relative thermograms" obtained with 5, 10 and 20°C/min as heating rates. The three curves show that three exotherms as a function of temperature for each heating rate. The first exotherm is located between 30 and 50°C (peak "1"), the second between 50 - 60°C and 95 - 105°C (peak "2") and the third between 95 - 105°C and polymer melting temperature (peak "3"). Peak "1" is too close to heating start-up and so can depend on thermal equilibrium of the system. Peak "3" is disturbed by melting endotherm differences between control P P and ozonized P P.

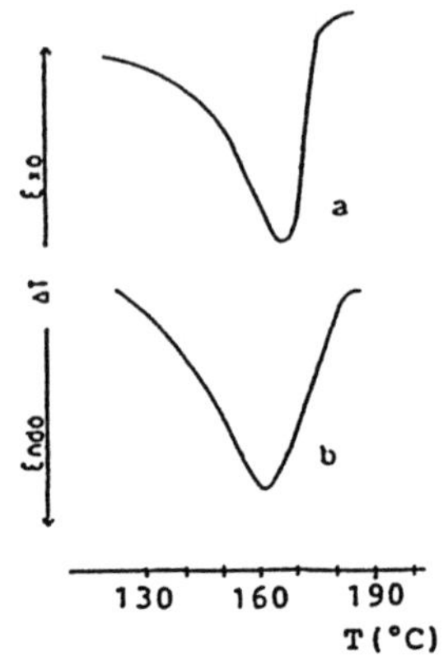

Graph 3 : *D S C "classical thermograms"*
Curve a - control P P
Curve b - ozonized P P

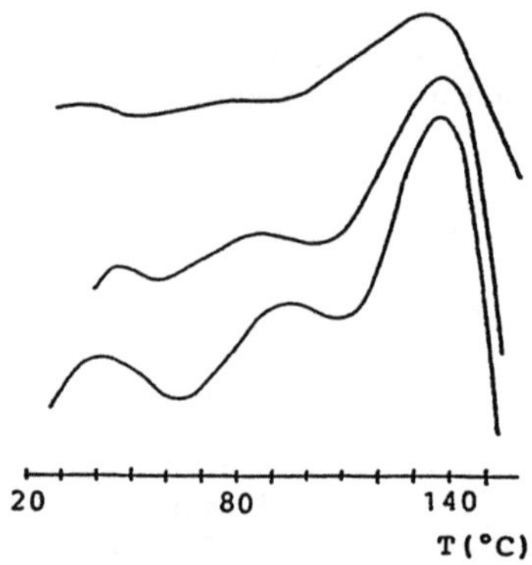

Graph 4 : *D S C "relative thermograms" ; upper curve : 5 °C / min, middle curve 10 °C / min, lower curve 20 °C / min.*
Refernce cell : control P P. Sample cell : ozonized P P

Starting (T_0) and maximum (T_m) temperatures for the three peaks described above are reported in Table 2.

V_T (°C/min)	T_{01} (°C)	T_{m1} (°C)	T_{02} (°C)	T_{m2} (°C)	T_{03} (°C)	T_{m3} (°C)
5	30	33	55-60	75-80	95	131
10	30	33	55	83	100	131
20	30	33	50	90	105	131

Table 2

DISCUSSION

E S R Results

Graphs 1 and 2 show the generated spin adduct nitroxide amount due to free radicals arising from the thermal decomposition of the peroxides and trapped by POBN and PBN. The chemical reactions can be summarized as follows :

$$\mathrm{ROOR} \xrightarrow{K_1} 2\,\mathrm{RO}^{\bullet}$$

$$\mathrm{ROOR} \xrightarrow{K_1} \mathrm{RO}^{\bullet} + {}^{\bullet}\mathrm{OH}$$

$$\mathrm{ST} + \mathrm{R}^{\bullet} \xrightarrow{K_2} \mathrm{SA}^{\bullet}\ \text{(Spin Adduct)}$$

It is well known that the slower reaction controls the kinetics of the process. An estimate rate of 10^{-3} min^{-1} for the rate constant of polypropylene peroxides thermal decomposition at 100°C is available (8, 9), while the rate of nitroxide formation (10) is much faster, as shown by Janzen and al. (11).

Isothermal curves (graphs 1 and 2) reach a constant value asymprotically whatever the nitrone molecule. On graph 5 the logarithm of initial rate is plotted versus reciprocal treatment temperatures. The experimental points are located on a straight line characteristic for an activated Arrhenius process

$$V_i = V_{i0} \exp(-E/RT)$$

where

V_{i0} is the pre-exponential factor
E the activation energy and
R the gas constant

The slope of the straight line gives the activation energy and the zero-ordinate the pre-exponential factor. Values obtained are :

$E = 102$ kJ/mol
$V_{i0} = 2,8 \,.\, 10^{12}$ min^{-1}

These values are in good agreement with thse of literaturre (8, 9) :

$E = 108 - 106$ kJ/mol
$V_{i0} = 2,3 - 3,2 \,.\, 10^{12}$ min^{-1}

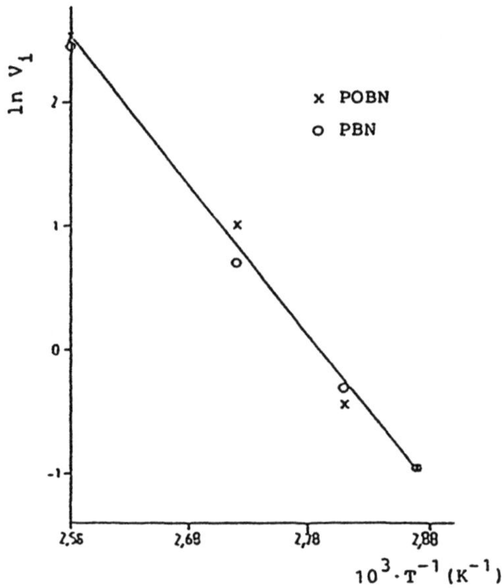

<u>Graph 5</u> : Arrhenius plot for thermal decomposition velocity

D S C Results

As far as relative thermograms are concerned, we took into account the D S C method described by Beyer for determination of activation energy of a peroxide dissociation in a polymer comple (12) :

$$2 \mathrm{Ln}\, T_{max} - \mathrm{Ln}\, a = E/RT_{max} + \mathrm{Ln}\, E/R - \mathrm{Ln}\, K_0$$

where

T_{max} is the maximum temperature of the exotherm
a the heating rate
E the activation energy
R the gas constant

A plot of (2 Ln T_{max} - Ln a) versus reciprocal maximum temperature allows activation energy determination. Becouse no E S R spectra are observed for peroxide thermal decomposition below 70°C, we choose to take into account the temperature of the maximum of the second peak T_{m2}.

Graph 6 shows results obtained in this case, and the activation energy value calculated form the slope of the straight line is :

$$E = 100 \text{ kJ/mol}$$

This value confirms E S R results and is in good agreement with values of the literature (8, 9).

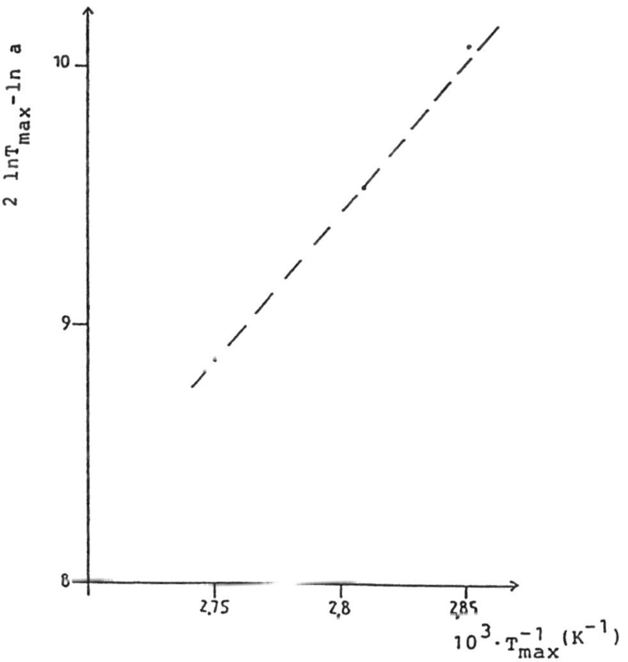

Graph 6 : *Beyer plot for thermal decomposition velocity*

CONCLUSION

The E S R data clearly show that the "Spin -Trapping" technique can be used successfully for quantitative measurements of polypropylene peroxides thermal decompositon in a polymer matrix. The actiation energy and rate constant values obtained by E S R are in good agreement with those found in lthe literature by other techniques. This technique permits to follow the thermal decomposition kinetics of the reactive groups at various temperatures. Moreover, the "relative thermogram" technique in D S C allows to obtain information about the peroxide decomposition exothermicity. The activation energies of the thermal decomposition of peroxides obtained by these two techniques are in good agreement (E = 100 kJ/mol). However, in the decompostion of peroxides more than one type of radicals was trapped. The identification of the chemical structure of these radicals is being attempted.

REFERENCES

(1) A. A. POPOV ET AL.
J. Polym. Sci., Phys. Ed., 21, 1017 (1983)

(2) G. SMZETS, G. WEINAND AND S. DEGUCHI
J. Polym Sci., Chem. Ed., 16, 3077 (1978)

(3) A. MICHEL, C. MONNET
Eur. Polym. J., 17, 1145 (1981)

(4) P. CITOVICKI, D. MIKULASOVA, V. CHRASTOVA AND G. BENC
Eur. Polym. J., 13, 655 (1986)

(5) V. VERNEY, A. MICHEL
Polym. Proc. Engn., 4, 335 (1986)

(6) B. CATOIRE, C. TASSIN, R. OUESLATI, C. PROST, V. VERNEY, A. MICHEL
Polym. Proc. Engn., 4, 271 (1986)

(7) R. M. JOHNSON, I. SIDDIQUI
"The determination of organic peroxides", Pergamon, London (1966)

(8) N. V. ZOLOTVA, E.T. DENISOV
J. Polym. Sci., A-1, 9, 3311 (1971)

(9) M. IRING
Polym. Degrad. and Stab. 3, 371 (1980-81)

(10) E. G. JANZEN, C. ANDERSON EVANS, Y. NISHI
J. Amer. Chem. Soc., 94, 8236 (1972)

(11) M. SCHMID, K. . INGOLD
J. Amer. Chem. Soc., 100, 2493 (1978)

(12) G. BEYER
Int. Polym. Sci. and Techn., 13, T/1 (1986)

This article was already published in 'Die Makromolekulare Chemie - Macromolecular Symposia" - Vol. 25 - January 1989

Spin-Trapping in Biological Systems

Claude CHOPARD

Laboratoire de Chimie et Biochimie Pharmacologiques et Toxicologiques, Unité associée au CNRS (URA 400), Université René Descartes, 45 rue des Saints-Pères - 75270 Paris Cedex 06, France

Spin trapping is a technique for detecting short-lived (reactive) free radicals. It consists in producing an ESR detectable (long-lived) radical by an addition reaction of the short-lived radical on a trapping agent. The spin trap is generally a nitrone or a nitroso compound. then the spin adduct an aminoxyl radical having at least a three lines spectrum. Nuclei having a magnetic moment can interact with the aminoxyl unpaired electron causing further hyperfine splitting whose constants can help to identify the trapped species.

The concentration of the components in biological systems are usually low, as the steady-state concentration of the short lived free radical. It is necessary to transform them in more stable free radicals in order to obtain their accumulation. In these systems, the free radicals formed are often generated by an oxidative metabolism; therefore many of them are O-centered radicals as HO°, O_2°-, RO°, ROO° ... and cannot be observed directly by EPR because of their relaxation time. The technique of spin-trapping is supposed to be able to overcome these difficulties and has been developed rapidly and very soon for biological systems: the first example is the study of the anaerobic reaction of Lipoxygenase-1 by De Groot et al in 1973[1]. This technique has generally been used to show that free radicals are end products or intermediates of several biochemical reactions, and to investigate the mechanism involved. As many papers appeared, but some of them with spurious conclusions, this methode and its limitations have been reviewed in recent years[2,3,4].

In this report which do not pretend to be exhaustive, we present some of these limitations and the hints which have been given in order to get valuable conclusions from experiments of spin-trapping In a second part we present a spin-trapping investigation of the radicals released during the dioxygenation of phenylhydrazones by soybean lipoxygenase-1 which has been performed recently in our laboratory[5,6].

I - ABOUT THE TECHNIQUE OF SPIN-TRAPPING:

The first aim, when a spectum has been obtained, consists in the assignement of the trapped radicals structure. The interpretation of spectra may be difficult when a mixture of several spin-adducts is formed. More often, one or two spin-adducts dominate in the experimental spectra. The the structure assignement of the trapped radical can be made with the help of the ESR parameters tables which have been published[7,8], and the data-bases now created[9,10]. Nevertheless, the very large solvent effects which have been observed on the magnitude of the splitting constants, may enhance considerably the assignement difficulty. A little prior knowledge of the reaction system under scrutinity would be helpfull. If the spin adduct is stable enough, it can be isolated (i.e. by HPLC), then analysed (i.e. by mass spectrometry). The right structure of the trapped radical can also be deduced from experiments using other spin traps, or generating the radical by an other route. It can also be determined from the effect of isotopic enrichment of the radical (^{13}C, ^{15}N or ^{17}O) or of the spin trap (^{13}C) on the spectrum.

B. Catoire (Editor)
Electron Spin Resonance (ESR) Applications in Organic and Bioorganic Materials

When investigating a biological system, before drawing definitive conclusions, it is first necessary to be sure that the spin-adduct is not an imposter. In fact, several routes can be involved, leading to this kind of species: as any chemical, the spin trap may react with any component of the system or be substrate of the enzyme(s) present in the medium; controls are needed to insure this point, in particular with the heat-inactivated enzyme(s). Impurities frequently found in commercial spin-traps (i.e.DMPO) give also rise to artefactual spectra.

One must also keep in mind the fact that, in spin-trapping investigations, the intensity of the ESR spectrum reflects a steady-state condition resulting of the balance between the rate of formation of the spin-adduct and the rate of its decay. The spectrum obtained depends on the stability of the spin-adduct by itself and on the fate of the spin-adduct in this complex medium. If the spin-adduct of the radical initially formed has a poor stability, it steady-state concentration might be too low to be detectable; its decomposition may lead to another more stable spin-adduct whith an observable spectrum. By example, it is now well known that DMPO°-OOH is not very stable, it decomposes giving DMPO°-OH wich is often the only adduct observed on the spectra. Therefore the system should be tested with differents spin-traps and ESR experiments have to be performed immediatly after mixing the reactants and also at various times.

From an other point of view, the spin-trap is a radical scavenger, which may alter the reaction sequence under investigation. This aspect must also be examined very carefully before drawing valuable conclusions.

The technique of spin-trapping is an interresting tool to investigate radical reactions in biological systems provided that its limitations are well understood. The study that we present above may be considered as a very simple example of a biological system as we have used a pure enzyme, moreover it needs nor cofactor nor coupled enzymatic system.

II - A SPIN TRAPPING INVESTIGATION: **The dioxygenation of phenylhydrazones catalysed by soybean lipoxygenase-1.**

The enzyme lipoxygenase (EC 1.13.11.12) is a non heme iron dioxygenase that catalyzes the dioxygenation of the 1(Z),4(Z)-pentadiene system of unsaturated fatty acids. In mammals, it converts arachidonic acid into 5(S)-hydroperoxy-6(E),8(Z),11(Z),14(Z)-eicosatetraenoic acid (5-HPETE), the precursor of leukotrienes, a family of compounds with various potent physiological activities [11,12]. Soybean lipoxygenase-1 (L-1) produces specifically 13(S)-hydroperoxy-9(Z),11(E)-octadecadienoic acid (13-HPOD) from linoleic acid and 15(S)-hydroperoxy-5,8,11,13-eicosatetraenoic acid (15-HPETE) from arachidonic acid (for reviews see Galliard & Chan, 1980, Vliegenthart & Veldink, 1982)[13,14]. Several catalytic cycles have been proposed for L-1[15,16,17], for the reticulocyte lipoxygenase[18] and for the 5-lipoxygenase of human polymorphonuclear leukocytes[19]. It is commonly admitted[14] that the iron of the enzyme shuttles between the III and II oxidation states (Scheme I, steps 1-4). In the first step, the ferric active form of the enzyme is reduced by linoleic acid. The radical R° reacts with a molecule of dioxygen to give the peroxyl radical ROO° (second step). In a third step, the iron (II) of the enzyme is oxidized by ROO° and the final hydroperoxide ROOH (13-HPOD) is released. Recently, an hydroperoxidase activity of L-1 without O_2 consumption has been described with several xenobiotics and 13-HPOD[20,21,22,23] that implies steps 1 and 5 of Scheme I (ROOH=13-HPOD).

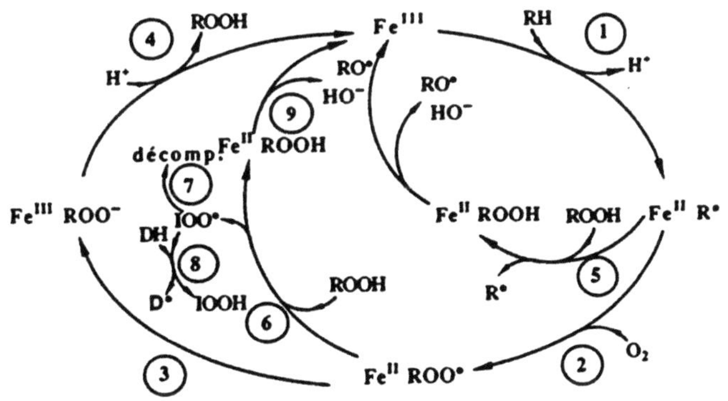

Scheme I

Catalytic cycle of soybean lipoxygenase-1 with the dioxygenase and the hydroperoxidase pathways for the various substrates and hydroperoxides. Substrates RH = linoleic acid, arachidonic acid, IH : phenylhydrazones Hydroperoxides ROOH = 13-HPOD, 9-HPOD, 15-HPETE or IOOH

We have previously shown that hexanal phenylhydrazone (I_1H) is a suicide-substrate of L-1, the α-azohydroperoxide I_1OOH being the inactivating species[24]. In order to determine the factors which could improve the efficiency of phenylhydrazones as inhibitors of L-1, we investigated in more details the various steps of the dioxygenation of four of them: hexanal phenylhydrazone I_1H, benzaldehyde phenylhydrazone I_2H, acetophenone phenylhydrazone I_3H, and 4-benzyloxyacetophenone phenylhydrazone I_4H. The four phenylhydrazones are potent inhibitors of L-1.

As previously found for I_1H, the hydrazones I_2H, I_3H and I_4H are substrates of L-1. Addition of one equivalent of IH to the ferric enzyme, under anaerobic conditions, leads to the reduction of the iron to its ferrous state, as shown by the complete disappearance of the g=6.1 EPR characteristic signal of L-1(Fe^{III}). Under aerobic conditions, dioxygen is taken up by the four phenylhydrazones. They differ by the velocity of this dioxygen consumption with values of initial velocities ($\mathcal{V}i$) comprised between 3.4 and 38 nmol. min^{-1} (experimental conditions: $[IH_0]$ = 100μM and [L-1] = 0.4μM in 1.2ml Tris-acetate buffer at pH9)

The amount of dioxygen consumed (O_{2c}) was found equal to that of converted IH. The theoretical amount of IOOH formed from the L-1 catalyzed dioxygenation of IH, according to steps 1-4 of Scheme I, should be equal to that of dioxygen consumed. But the amount of IOOH detected ($IOOH_d$) in the reaction medium was always less than that expected on the basis of dioxygen uptake, independently of the molar excess of IH over L-1 (IH_0/L-1). $IOOH_d$ increases with IH_0/L-1 and O_{2c}, but the ratio of $IOOH_d$ to O_{2c} reaches only 32% for I_1H, 5% for I_2H, 18% for I_3H, and 86% for I_4H when IH_0/L-1= 500.

The "missing" IOOH ($IOOH_m$, per mole of L-1) is not due to its unstability, because the authentic α-azohydroperoxides were found to be stable in the buffer solutions used, during the time of the experiments in the absence of L-1 or in the presence of boiled L-1.

In another set of experiments, we have investigated the influence of added 13-HPOD on the L-1 catalyzed transformation of the four hydrazones. 13-HPOD, in a 1.2:1 ratio versus IH_0, was added at the beginning of the reactions. The amount of dioxygen consumed was still equal to that of converted phenylhydrazone. Moreover: i the addition of 13-HPOD increases both O_{2c} and therefore the number of turnovers before L-1 inactivation, except for I_4H, and the rate of dioxygen consumption with v_i ranging from 88 to 440 nmol. min^{-1} instead of 3.4 to 38 nmol. min^{-1}; this acceleration is larger in the cases of I_2H and I_3H compared to those of I_1H and I_4H; ii $IOOH_d$ increases as well except for I_4H, and it remains smaller than O_{2c}; iii 13-HPOD is consumed; 13-$HPOD_c$, is larger for I_2H and I_3H than for I_1H and I_4H, corresponding respectively to 100 and 67% of the amount of converted IH compared to 35 and 26%, when IH_0/L-1= 500. Similar effects have been observed with other biological relevant hydroperoxides: 15-HPETE and 9-HPOD.

The "missing" IOOH and the effect of added 13-HPOD suggested a branched pathway for the L-1 Fe^{II} IOO° species (Scheme I): either the 3 + 4 dioxygenase route that leads to IOOH, or the hydroperoxidase route 6 + 9 that releases IOO° in the medium and regenerates the active Fe^{III} enzyme by IOOH oxidation of the Fe^{II} species. The released IOO° can give IOOH (step 8) or decomposes (step 7). Steps 6 and 9 can be promoted by the addition of a fatty acid hydroperoxide and the rate of the overall oxidation depends both on the nature of IH and on that of ROOH actually involved in steps 6 and 9. The difference observed between the hydrazones reflects the contribution of the peroxidase pathway (step 6) and of the decomposition step 7 in the overall process.

In order to confirm this "leakage" of IOO° from the active site, we try to trap this species. We chose 5,5-dimethyl-1-pyrroline-N-oxide (DMPO) and α-4-pyridyl-1-oxide-N-*tert*-butylnitrone (4-POBN) which are water-soluble and also 2-methyl 2-nitroso propane (MNP) known to react with C° radicals. We checked that in our conditions they gave no signals with L-1 alone (Figures 1a and 2a), and no (Figure 1b) or very small (Figure 2b) signals with the phenylhydrazones alone. The EPR experiments were performed with 70-90 µM L-1, 100mM DMPO or 50mM 4-POBN or 100mM MNP, and 1mM of the various phenylhydrazones (IH_0/L-1 = 12.5) except for I_4H which was 0.5mM (IH_0/L-1 = 7.5).

We have first looked at the effect of these spin-traps on the rate of dioxygenation of the phenylhydrazones (Table 1). MNP have no effect at all. With DMPO and POBN, here again the amount of dioxygen consumed was found equal to that of phenylhydrazone converted; the rate of dioxygen consumption is greatly reduced and the phenylhydrazones are not completely converted even when L-1 activity is remaining, except for I_4H. It is noteworthy that the spin-traps do not disturb the dioxygenation of linoleic acid in the same conditions and therefore exhibit a lower affinity for the L-1 active site.

As the K_I of the four IHs (0.2, 0.4, 2.4 and 5.6µM for I_1H, I_2H, I_3H and I_4H respectively) are lower than the Km found for linoleic acid (24µM), this rules out a competitive inhibition of the dioxygenation of IH by the spin-traps. The inhibition observed must be due to the trapping of radical species in the medium, which are involved in the dioxygenation process. As the amount of dioxygen consumed is still equal to that of the hydrazone converted, the radical trapped cannot be I° released by step 5 but could be IOO°. It is noticeable that the spin traps have no effect on the reaction performed in the presence of 13-HPOD.

IH_0/L-1 [a],	Spin-trap and its concentration	rate of O_2 consumption (%) compared to those without	O_2 consumption (%) of the experiments spin-trap	Remaining activity(%)
I_1H_0/L-1=15	MNP 20mM	100	100	0
I_1H_0/L-1=15	DMPO 20mM	<3	5	92
I_1H_0/L-1=15	4-POBN 10mM	54	95	0
I_1H_0/L-1=7.5		9	19	45
I_1H_0/L-1=15	4-POBN 20mM	11	23	22
I_2H_0/L-1=15	4-POBN 10mM	<3	<5	85
I_3H_0/L-1=15		<3	<5	85
I_4H_0/L-1=7.5	4-POBN 10mM	67	100	21
LH_0/L-1=15	4-POBN 20mM	100	100	100

a Concentrations used for I_1H, I_2H, I_3H and LH: 200µM and for I_1H and I_4H at IH_0/L-1 = 7.5: 100µM.

Table 1

Effect of the spin-traps MNP, DMPO and 4-POBN on the rate and amount of dioxygen consumption, compared to those obtained without spin-trap for the L-1 catalyzed dioxygenation of the four phenylhydrazones[a] and linoleic acid (LH).

No EPR spectrum is obtained when MNP is used as the spin-trap. This confirms that I° radicals are not released, then that step 5 do not take place.

The EPR spectra observed for the dioxygenation reaction of I_1H catalyzed by L-1, in the presence of DMPO and 4-POBN are shown on Figures 1c and 2c. They are identical in each case to those obtained with I_2H, I_3H, and I_4H (data not shown). The four phenylhydrazones give the same spin-adducts exhibiting the following EPR parameters: a_N= 15.9G, a_H= 24.7G for DMPO (Figure 1c) and a_N= 15.7G, a_H= 3.3G for 4-POBN (Figure 2c). The intensity of the EPR signals depends on the phenylhydrazone, increases with the concentrations of L-1 and phenylhydrazone. It increases also with the length of aerobic incubation before recording the spectra (checked from 15 to 120min) and decreased when the sample was kept in the flat EPR cell which progressively became anaerobic. Identical spectra are observed when the spin trap is added to L-1 before the hydrazone or when the enzyme is added to the mixture of spin trap and hydrazone. When one equivalent of hydrazone is added to L-1 Fe^{III} (prepared free from excess of 13-HPOD) the same spectra are obtained with lower intensity.

The EPR parameters of the common spin-adduct formed in the various experiments suggested the trapping of a phenyl radical[7]. However EPR parameters are dependent on the medium and this prompted us to run two control experiments. We have used the autoxidation of phenylhydrazine initiated by $CuCl_2$[25] and the decomposition of methyl phenylazoformate[26] which are known to give phenyl radicals via the formation and decomposition of phenyldiazene. In our Tris-acetate pH9 buffer, the phenyl-DMPO adduct has EPR parameters a_N= 16.0G, a_H= 24.7G

(Figures 1d and 1e) and the phenyl-4-POBN adduct has EPR parameters a_N= 15.7G and a_H= 3.3G (Figures 2d and 2e) identical to those observed for the L-1 reactions with I_1H.

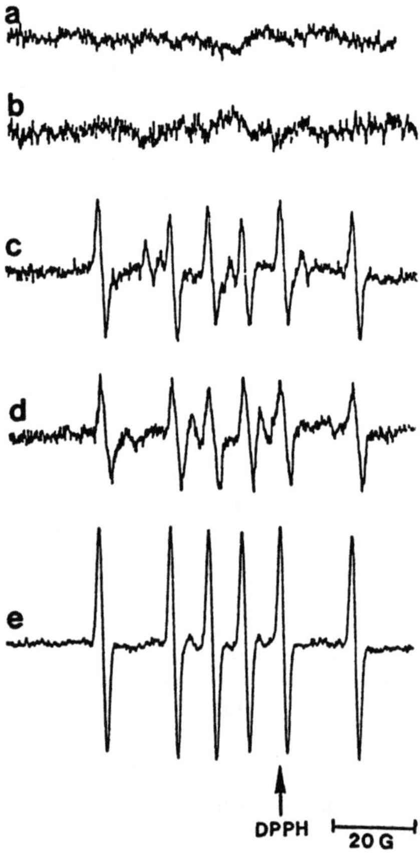

Figures 1

EPR spectra of the trapped species formed during the oxidation of hexanal phenylhydrazone (I_1H) catalyzed by L-1 in the presence of DMPO (Figure 1) All the spectra were recorded after 15-30 minutes aerobic incubations. Instrumental conditions: field set 3431G, field scan 100G, modulation frequency 2Gpp, receiver gain 10^6 (DMPO), 10^3-10^6 (4-POBN), time constant 200ms, scan time 200s, microwave frequency 9.55GHz, microwave power 10 mW, DPPH: diphenylpicrylhydrazyl (g=2.004 G).

a) 80µM L-1, 100mM DMPO
b) 1mM I_1H, 100mM DMPO
c) 80µM L-1, 1mM IH, 100mM DMPO
d) 1mM phenylhydrazine, 17µM $CuCl_2$, 100mM DMPO
e) 100µM methyl phenylazoformate, 100mM DMPO.

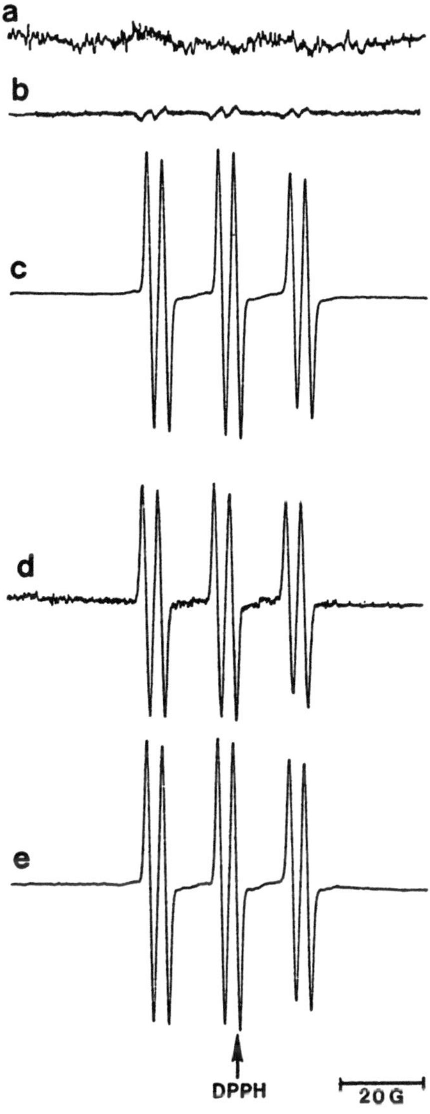

Figures 2

EPR spectra of the trapped species formed during the oxidation of hexanal phenylhydrazone (I_1H) catalyzed by L-1 in the presence of 4-POBN. All the spectra were recorded after 15-30 minutes aerobic incubation.
Instrumental conditions are the same as those described for figure 1.

a) 80μM L-1, 50mM 4-POBN, gain=10^6,
b) 1mM I_1H, 50mM 4-POBN, gain=1.25×10^5
c) 80μM L-1, 1mM I_1H, 50mM 4-POBN, gain=4×10^4
d) 1mM phenylhydrazine, 17μM $CuCl_2$, 50mM 4-POBN
e) 100μM methyl phenylazoformate, 50mM 4-POBN, gain=1.25×10^5

It has been reported that the spin adducts of peroxyl radicals are unstable in basic aqueous solution[28,29]. Therefore the trapped phenyl radical of our experiments could be a secondary adduct. We checked this point as follows: I_1H 1mM in toluene was autoxidized under a dioxygen atmosphere[27]. The 4-POBN spin adduct gave the following EPR parameters in toluene: a_N=14.8G, a_H=1.7G (Figure 3a).

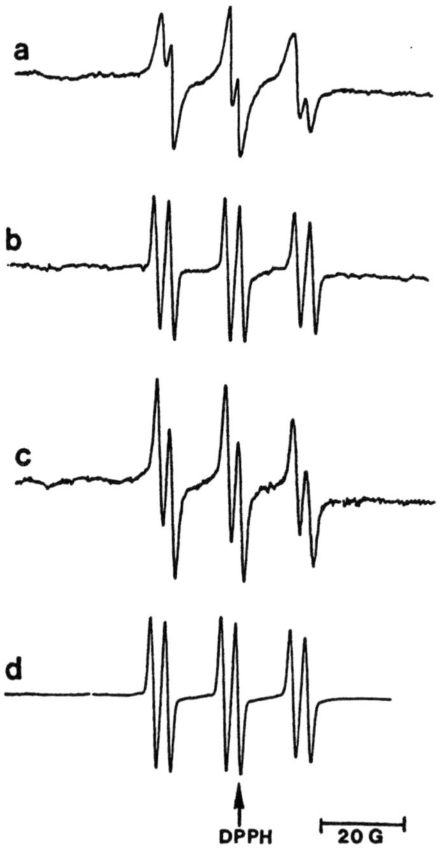

Figure 3

EPR spectra of the 4-POBN trapped species formed during the autoxidation of I_1H in toluene (3a), further extracted by Tris-acetate buffer pH9 (3b) or during the autoxidation of phenylhydrazine in toluene (3c), further extracted by Tris-acetate buffer pH9 (3d).
Instrumental conditions are the same as those described for figure 1.
a) 1mM I_1H in toluene, O_2, 50mM 4-POBN, gain=$3.2x10^5$
b) (a) extracted with the same volume of Tris-acetate buffer, gain=$3.2x10^5$
c) 1mM phenylhydrazine in toluene, O_2, 50mM 4-POBN, gain=$3.2x10^5$
d) (c) extracted with the same volume of Tris-acetate buffer, gain=$3.2x10^4$.

Extraction of the spin adduct and excess 4-POBN with the Tris buffer used in our experiments, gave the previously characterized phenyl adduct with a_N=15.7G and a_H=3.3G (Figure 3b). The autoxidation of phenylhydrazine in toluene [28] gave a phenyl-4-POBN adduct with the EPR parameters a_N=15.2G and a_H=2.4G (Figure 3c). This spectrum is quite different from that obtained during autoxidation of I_1H in the same solvent. Extraction of the sample with the Tris-buffer gave again the typical parameters of the phenyl-4-POBN adduct in this medium (Figure 3d).

The species trapped by both DMPO and 4-POBN is actually the phenyl radical. There was no evidence for the trapping of a hydrazonyl radical I°, either bound through nitrogen or carbon. The phenyl-adduct could be due to the decomposition of a former IOO°-adduct[29,30]. We have shown that the autoxidation of I_1H in toluene which is a preparative reaction for I_1OOH[27] and therefore likely generates $I_1OO°$ in the medium, gave a 4-POBN spin-adduct different from that of the phenyl radical formed in toluene from the autoxidation of phenylhydrazine[28]. The presumed $I_1OO°$-4-POBN adduct gave the phenyl-adduct after extraction with the pH9 Tris-acetate buffer used for the L-1 catalyzed reactions. This shows that the phenyl radicals, trapped during the L-1 catalyzed dioxygenation, can be due to the decomposition of the former IOO°-adducts. However, one cannot exclude the trapping of some phenyl radicals derived from the decomposition of IO° formed at step 9, but this should not affect the rate and amount of O_2 consumption. That DMPO and 4-POBN indeed trap IOO° is supported by their inhibition of the dioxygenation of IH.

The trapping of IOO° agrees with the relation that exists between the efficiency of the 4-POBN inhibition and the contribution of the hydroperoxidase pathway to the overall oxidation process. For example, the 4-POBN inhibition is very large for I_2H whose dioxygenation gives only 5% of I_2OOH_d vs. O_{2c} (IH_o/L-1=500), which shows that it is strongly dependent on step 6 involving I_2OOH formed from $I_2OO°$ by steps 6+8. On the contrary, the inhibition is small for I_4H whose dioxygenation gives 86% of I_4OOH_d vs. O_{2c} showing that steps 6+9 have a small contribution to the overall oxidation process which does not depend primarily on I_4OOH formed by steps 6+8.

Our results suggest that when the ROO° species is not properly suited for the electron transfer step 3: it can "leak" out of the active site, the latter requiring the oxidation step 9 to recover its catalytic activity. This could be related to several reports of the requirement of the hydroperoxide derivatives of fatty acids to sustain the catalytic activity of soybean and reticulocyte lipoxygenases[31,32].

As far as the search for better phenylhydrazone inhibitors is concerned, our results show that one should look for compounds with structures favoring step 3, like I_4H, but also with electronic properties leading to a very unstable IOOH generated at step 4 that would readily oxidize the active site methionine[24].

BIBLIOGRAPHY

1 De Groot, J. J. C., Garsen, G. J., Veldink, G. A., Vliegenthart, J. F. G. and Bolding, J., (1973) *Biochim. Biophys. Acta* , **326**, 279-284.

2 Janzen, E, G.,(1980) in *Free Radicals in Biologiy,* Vol IV (Pryor, W. A., Ed.) Academic Press, pp 115-154

3 Mason, R. P.,(1984) in *Spin Labelling in Pharmacology,* Academic Press, pp 87-129

4 Pou, S., Hassett, D. J., Britigan, B. E., Cohen, M. S. and Rosen, G. M.,(1989), *Anal. Biochem.* **177**, 1-6.

5 Bombard, S., Thesis, Université René Descartes Paris (1990).

6 Bombard, S., Chopard, C., Nam , N. H. and Chottard, J-C.,(in preparation)

7 Buettner, G. R., (1987), *Free Rad. Biol. Med.*, **3**, 259-303.

8 Perkins, M. J., (1980), in *Adv. Phys. Org. Chem.*, Vol. **7**, pp 1-64.

9 Li, A. S. W., Cummings, K. B., Roethling, H. P., Buettner, G. R. and Chignell, C. F., (1988), *J. Mag. Res.*, **79**, 140-142.

10 Dubose, C. M., Rehorek, D., Oehler, U. M. and Janzen, E. G., (1988), *Free Rad. Biol. Med.*, **5**, 55-56.

11 Samuelsson, B., (1983) *Science* ,,**220**, 568-575.

12 Samuelsson, B., Dahlen, S. E., Lindgren, J. A., Rouzer, C. A. & Serhan, C. N., (1987) *Science* , **237**, 1171-1176.

13 Galliard, T. & Chan, H. W. S., (1980) *The Biochemistry of Plants,* Vol.4, Academic Press, New York, pp 131-160.

14 Vliegenthart, J. F. G. & Veldink, G. A. (1982) in *Free Radicals in Biology,* Vol V, (Pryor, W. A., Ed.), Academic Press, New York, pp 29-64.

15 De Groot, J. J. C., Veldink, G. A., Vliegenthart, J. F. G., Bolding, J., Wever, R. & Van Gelder, B. F., (1975) *Biochim. Biophys. Acta* , **377**, 71-79

16 Egmond, M. R., Brunori, M. & Fasella, P. M., (1976) *Eur. J. Biochem.*, **61**, 93-100.

17 Lands, W. E. M., (1984) *Prostaglandins, Leukotrienes Med.*, **13**, 35-46.

18 Ludwig, P., Holzhütter, H. , Colosimo, A., Silvestrini, M. C., Schewe, T. & Rapoport S. M., (1987) *Eur. J. Biochem.*, **168**, 325-337..

19 Hatzelmann, A., Schatz, M. & Ullrich, V. (1989) *Eur. J. Biochem.*, **180**, 527-533.

20 Clapp, C. H., Banerjee, A. & Rotenberg, S. A. (1985) *Biochemistry,* **24**, 1826-1830.

21 Kemal, C., Louis-Flamberg, P., Krupinski-Olsen, R. & Shorter, A. L. (1987) *Biochemistry* , **26**, 7064-7072.

22 Mansuy, D., Cucurou, C., Biatry, B. & Battioni, J. P. (1988) *Biochem. Biophys. Res. Commun.* , **151**, 339-346.

23 Kulkarni, A. P. & Cook, D. C. (1988) *Res. Commun. Chem. Pathol. Pharm.* , **61**, 305-314.

24 Galey, J. B., Bombard, S., Chopard, C., Girerd, J. J., Lederer, F., Do-Cao-Thang, Nguyen-Hoang-Nam, Mansuy, D. & Chottard, J.C. (1988) *Biochemistry,* **27**, 1058-1066.

25 Hill, H. A. O & Thornalley, P. J . (1981) *FEBS Lett.*, **125**, 235-238.

26 Huang, P. C. & Kosower E. M. (1968) *J. Am. Chem. Soc.*, **90**, 2354-2361.

27 Yao, H. C. & Resnick, P. (1965) *J. Org. Chem.*, **30**, 2832-2834.

28 Hardie, R. L. & Thomson, R. H. (1957) *J. Chem. Soc.* , 2512-2518.

29 Rosen, G. M. & Rauckman, E. J. (1980) *Mol. Pharmacol.*, **17**, 233-238.

30 Davies, M. J., (1988), *Biochim. Biophys. Acta,* **964**, 28-35.

31 Smith, W. L. & Lands, W. E. M., (1972) *J. Biol. Chem.*, **247**, 1038-1047.

32 Kühn, R., Stender, H., Schewe, T., Lankin, V. Z., Nekrasov, A. & Rapoport, S. M. , (1986) *FEBS* , **203**, 247-252.

Electron Resonance Identificaton of Ionized Foodstuffs

J. RAFFI and J-P. AGNEL

Département de Physiologie du Végétal et Ecosystèmes

CEN Cadarache, 13108 Saint-Paul-lez-Durance, France

1- INTRODUCTION

Among the approaches used for reduction of food losses and for improvement of their hygienic quality, ionizing treatments are the most recent ones. In 1980, the Joint FAO/IAEA/WHO Expert Committee [1] approved irradiation up to an absorbed dose of 10 kGy. As a consequence, the radiation treatment of different foods is now legally accepted in many countries, although it is still prohibited in some others. Regulatory authorities in all countries appear to be interested in having simple and reliable methods to detect foods in the trade [2] and, consequently, to check on compliance with labelling regulations.

Let us first remember that an irradiation dose of 1 gray corresponds to the absorption of an energy of 1 joule per kilogram of foodstuff; the consequences of the irradiation are always linked to the dose which varies, depending on the purpose of the treatment, from about 0.1 kGy to 50 kGy. A lot of methods were discussed several times [2, 3] and, during the two last years, many data were published using the Electron Spin Resonance (ESR) technique and leading to the very first test [4]. The purpose of this paper is to show how ESR can be used to identify whether or not the food have been irradiated.

B. Catoire (Editor)
Electron Spin Resonance (ESR) Applications
in Organic and Bioorganic Materials

2- CONDITIONS REQUIRED TO USE ESR AS AN IDENTIFICATION TEST

ESR is a spectroscopic method allowing the obervation of alone electrons, wich happens in three cases:

- semi-conductors which is never the case in foods
- ions from transition elements which often lead to signals present in the reference sample as a lot of food enzymes are working with such ions;
- radicals and, especially, those induced by irradiation.

ESR can be used as an identification test if [5-7]:

- the radicals are quite stable with regard to the maximal time of commercial storage of the considered food; this is only the true in solid and dry state, where the reactivity of the radicals with each other or with water is low;
- the relative signals must be clearly distinguishable from the ones of the reference sample as we never have it in commercial conditions.
- the signal must be strictly constant if we also require an estimation of the initial dose.

3- DETECTION OF MEAT AND FISH

Studying the ESR signals induced in bones is not new as yet done for post-irradiation dosimetry [8] and for dating of bones and shells [9]. In this datum, we shall not develop details about the recording conditions as two other data will speak about it [10, 11]. We only point out the main results (Fig. 1):

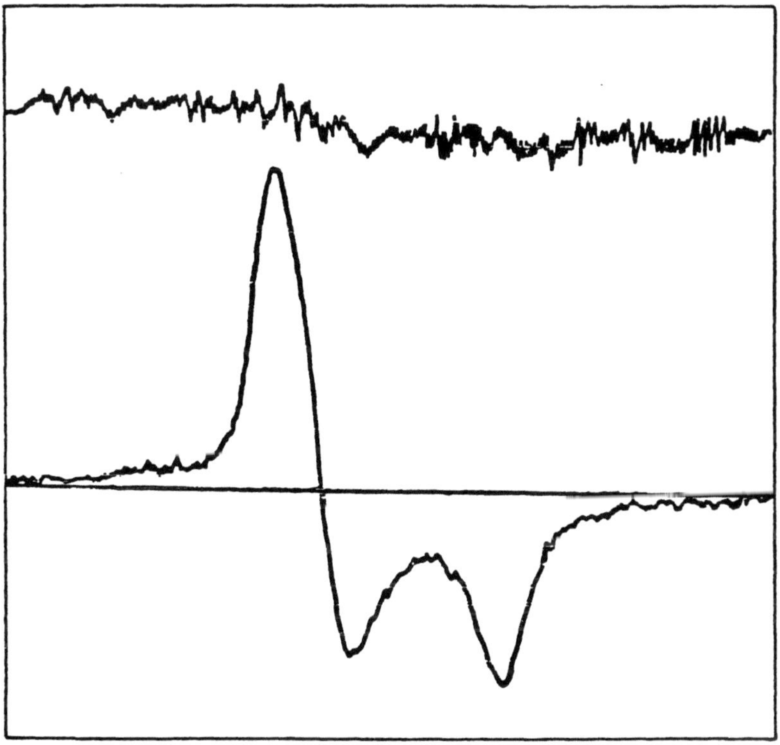

Figure 1: ESR spectrum of bones from non irradiated (top) and irradiated (low) sardin (Field: 350 ± 35 mT).

- in the unirradiated samples, there is a symmetric absorption with a g-factor ~ 2.0044 and a peak-to-peak linewidth (Δ H) of ~ .6 mT. In the case of frozen frog bones, it is quite only due to the marrow [5];

- an asymmetric absorption characterized by

$g(\perp)$ ~ 2.0030 - 2.0033

and $g(//)$ ~ 1.9969 - 1.9975

which is only observed in irradiated samples and due to the hydroxyapatite [$Ca_{10}(PO_4)_6(OH)_2$] : frog, lamb, rabbit and

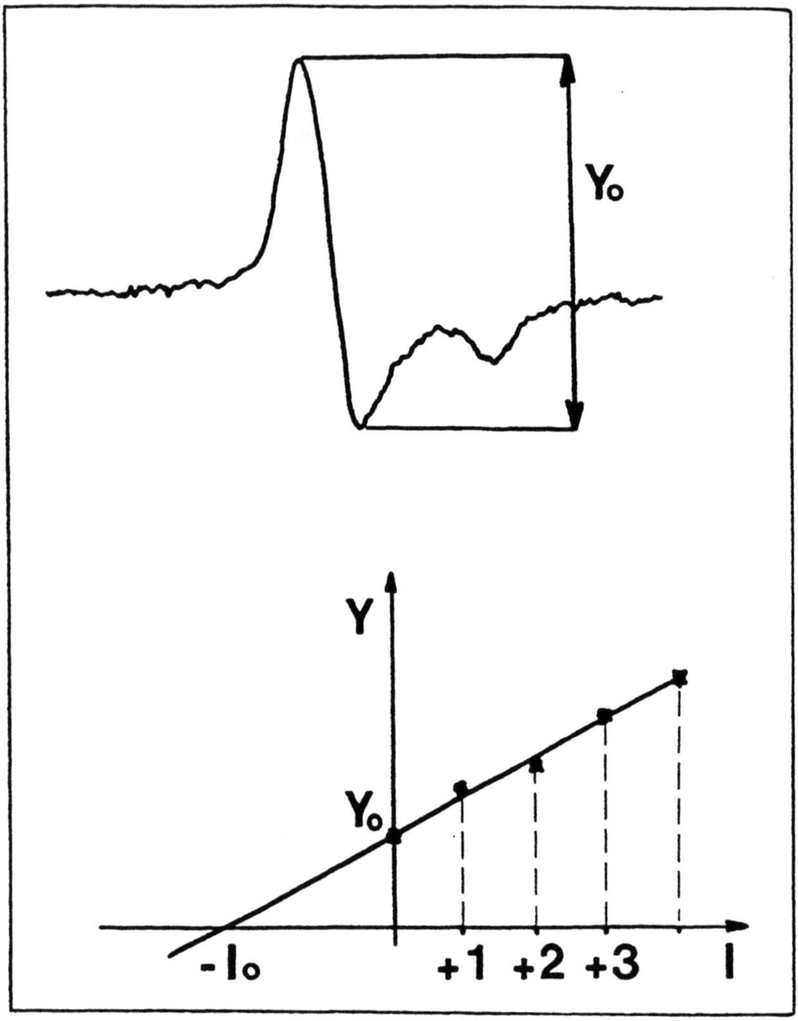

Figure 2: Determination of the initial dose delivered to a bone: the heigth of the successive signals is reported on a graphic in order to reach the initial dose by extrapolation.

poultry bones, sardine, brown trout, eel and pike-trech bones, scales or jaw [5, 12-15].

- Since the intensity of signal height is proportional to the dose (up to 14 kGy in the case of frog legs), it is possible to obtain an estimate of it in case of meat bones; this can be overcome if a "standard addition" technique is employed: the sample would be re-irradiated several times with its ESR signal recorded after every re-irradiation (Fig. 2). Such a technique has been used for the dating of fossil shells [16]; because of heterogeneity in the structure of the bones, it only allows an estimation of the initial dose [12, 14-16], which is sufficient enough to check if the purpose of the irradiation was, or not, the decrease of the microbiological dose (range 4-10 kGy).

3- DETECTION OF FRUITS

As the water content of fruits is relatively high, the radicals induced in the pulp are not stable; however ESR can be use in the case of high dry matter components such as achenes, pips or stones. The method leaded, in the case of strawberries, to the first test [45], and has been recently extended to other fruits [17]: an ESR spectrum of achenes, pips or stone, is due to two or three different signals (Fig. 3):

- a six line signal (A) due to Mn^{2+}, both present (or missing) in irradiated and non irradiated samples [4, 13, 18];

- a central single line (B), also present in the non irradiated sample, which increases with the irradiation dose, but varies to a large extent with the water content; it

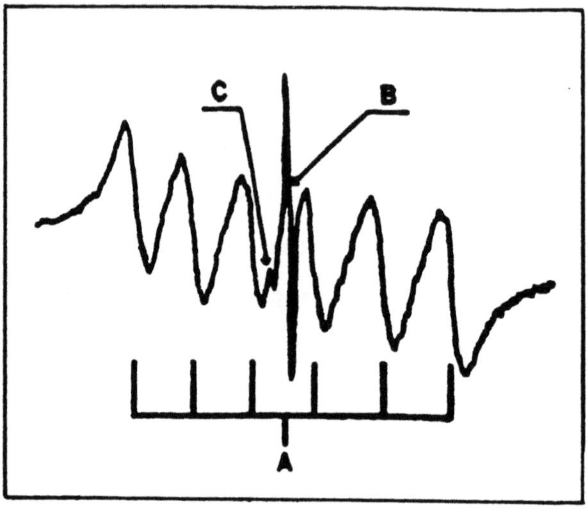

Figure 3: ESR spectrum of irradiated raspberry akenes (field: 350 ± 80 mT; modulation: .5 mT)

may be related to a quinone radical [17];

- a triplet (C), due to a cellulose radical [17], the left line of which can be used as a test of irradiation, at least for berries and dried fruits. For the other fruits, experiments have to be carried out in order to check if the "stability" of this particular line is longer than the maximal commercial storage time of the fruit.

4- OTHER PRODUCTS

A lot of other ESR signals can be observed in foodstuffs: egg, snail or sea shells [4, 16], cuticules of crustacea [19], dried spices [20] or cereals [6]; always, the problem is to check if the "characteristic" signal is observed longer than the usual commercial storage time; for instance, in the case of wheat, the signal, due to starch radicals, is distinguishable from the reference for more than one one year after irradiation; however, these long lived species are not enough stable with regard to the possible 4-5 years storage.

5- CONCLUSION

Even if the use of ESR spectroscopy does not allow a test available for all foodstuffs, it leads to a proof in meat and fish bones, and in some fruits. It would be the first "general" method recognized by the health authorities. A small co-trial was yet organised on meat bones [21] with the participation of 4 laboratories (Paterson Institute and University College of Cardiff in U-K, NST in the USA and CEN Cadarache in France) but, in 1990, a more important intercomparison will be organized on meat and fish bones and fruits by our laboratory with 20-30 other ones with the sponsoring of the Community Bureau of Reference (BCR) of the Commission of the European Communities. This intercomparison is part of a more general programme on identification methods of irradiated food, the organization of which I am in charge, with the help of the BCR.

Acknowledgments: this work was partly sponsored by the International Atomic Energy Agency (Research Agreement N° 5154/CF) and the French Research Ministery (Grant N°88.G.104). We also used the informations given by the participants of the two last BCR meetings organized in Brussells (3-4 October 1989) and Cadarache (13-15 February 1990).

REFERENCES

1- Joint FAO/IAEA/WHO Expert Committee, Wholesomeness of Irradiated Food, WHO, Geneva, 1981, N°659.

2- FAO/IAEA Research Coordination Meeting on Food Irradiation Programme for the Middle East and European Countries, Wageningen, 17-21 April 1989.

3- H. DELINCEE, D. EHLERMANN, Recent advances in the identification of irradiated food, Radiat. Phys. Chem.,1989, 34, 877-890.

4- J. RAFFI, J-P. AGNEL, L. BUSCARLET, C. MARTIN, ESR Identification of Irradiated Strawberries, J. Chem. Soc. Farad. Trans. I, 1988, 84, 3359-3362.

5- J. RAFFI, J-P. AGNEL, Identification par RPE d'aliments ionisés, Ann. Fals. Exp. Chim., 1989, 82, 279-287.

6- J. RAFFI, J-P. AGNEL, S-R. KASSIS, Identification par RPE de céréales irradiées, Sci. Alim., Paris, 1987, 7, 657-663.

7- A. SWALLOW, Can we tell if our food has been irradiated? Chem. Britain, 1988, 24, 102-103.

8- M. IKEYA, T. MIKI, ESR dating of Animal and Human Bones, Science, 1980, 207, 977-979.

9- M. IKEYA, K. OHMURA, Dating of fossil shells with ESR, J. Geology, 1981, 89, 247-251.

10- A. M. ROSSI, G. POUPEAU, O. CHAIX, J. RAFFI, J-P. AGNEL, A. JEUNET, Paramagnetic species induced in bioapatites during ionizing treatment of food, this issue.

11- M. H. STEVENSON, R. GRAY, Some factors influencing the ESR signal from irradiated poultry products containing bone, this issue.

12- M. DESROSIERS, M. SIMIC, Post-irradiation dosimetry of meat by ESR spectroscopy of bones, J. Agric. Food Chem., 1988, 36, 601-603.

13- N. DODD, J. LEA, A. SWALLOW, The ESR detection of irradiated food, Appl. Radiat. Isot., 1989, 40, 1211-1214.

14- R. GRAY, M. H. STEVENSON, Detection of irradiated deboned turkey meat using ESR spectroscopy, Radiat. Phys. Chem., 1989, 34, 899-902.

15- J. RAFFI, J. EVANS, J-P. AGNEL, C. ROWLANDS, G. LESGARDS ESR Analysis of Irradiated Frog Legs and Fishes, Appl. Radiat. Isot., 1989, 40, 1215-1218.

16- M. DESROSIERS, Gamma-irradiated seafoods: identification and dosimetry by EPR spectroscopy, J. Agric. Fodd Chem., 1989, 37, 96-101.

17- J. RAFFI, J-P. AGNEL, ESR Identification of Irradiated Fruits, Radiat. Phys. Chem., 1989, 34, 891-894.

18- M. DESROSIERS, W. McLAUGHIN, Examination of gamma-irradiated fruits and vegetables by ESR spectroscopy,Radiat. Phys. Chem., 1989, 34, 895-898.

19- J. RAFFI, J-P. AGNEL, ESR of Crustacea as an example of identification of ionized food, Sci. Alim., Paris, in press.

20- G. YANG, M. MOSSOBA, U. MERIN, I. ROSENTHAL, An EPR study of free radicals generated by gamma radiation of dried spices and spray-dried fruit powders, J. Food Quality, 1987, 10, 287-294.

21- M. DESROSIERS, N. DODD, J. EVANS, C. ROWLANDS, J. RAFFI, J-P. AGNEL, Co-trial on Identification of gamma-irradiated boned food, in preparation.

Note: Three reports will be published by the BCR (Commission of the European Communities presenting the main results of the Cadarache meeting (end of 1990) and on the two intercomparisons organized on Electron Spin Resonance and Thermoluminescence (both in 1991).

Some Factors Influencing the ESR Signal from Irradiated Poultry Products Containing Bone

(1) (2) M. Hilary STEVENSON and (1) Richard GRAY

(1) Food and Agricultural Chemistry Research Division
Departement of Agriculture for Nothern Ireland

(2) The Queen's University of Belfast
Newforge Lane
Belfast BT9 5PX
Northern Ireland, UNITED KINGDOM

The process of irradiation can be controled by good management at the irradiation facility and dosimeters can be used routinely to measure the dose received by food. Nevertheless, it is generally accepted that the development of a test to detect irradiated food would help to facilitate international trade in these foods and would enhance consumer confidence in the existing control procedures.

A number of different approaches to the problem are being investigated and electron spin resonance (ESR) spectroscopy offers potential as a method for the detection of irradiated food containing bone. The radiation-induced free radicals which are trapped in the crystal lattice of bone give an ESR signal quite different in shape to that from non-irradiated bone (Figure 1). As well as using the techniques as a qualitative test to detect irradiated food, it was of interest to establish the feasibility of using the technique to quantify the dose received by irradiated bone.

A series of experiments have been carried out in order to investigate the effect of different variables on the strength of this radiation-induced signal
These have included :

(a) method of sample preparation

(b) irradiation dose

(c) length and temperature of storage

(d) post-irradiation cooking

(e) age of bird

(f) site of bone within the carcass.

B. Catoire (Editor)
Electron Spin Resonance (ESR) Applications
in Organic and Bioorganic Materials

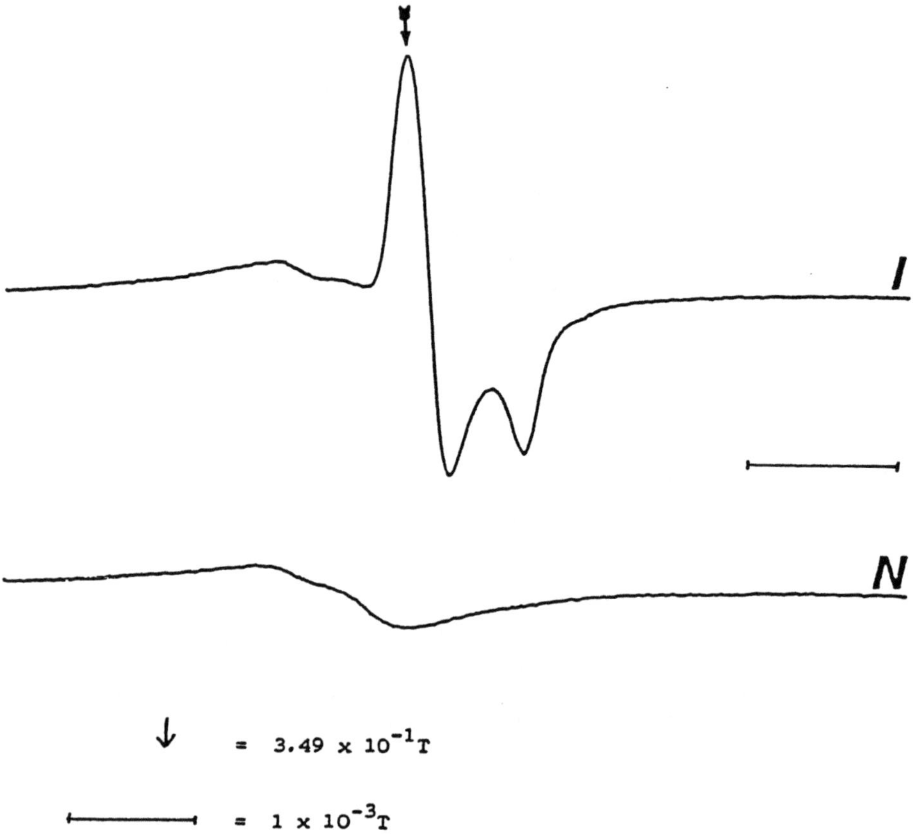

Figure 1: ESR signal from irradiated (I) and non-irradiated (N) chicken bone

A number of sample preparation methods have been investigated (1). It was concluded that freeze drying and grinding provided the most homogeneous sample and gave a signal with minimal interferences caused by the method of preparation. This method of sample preparation was then used to investigate the effect of the above variables on the ESR signal strength. The intensity of the ESR signal was estimated by double integration of the radiation-induced signal after substraction of the signal from the non-irradiated bone and correcting the value obtained to a standard dry weight of sample in the "active length" of the ESR tube (1, 2).

Using chicken bones (tibia), it has been shown (Table 1) that there is a linear relationship between ESR signal strength and irradiation doses between 2.5 and 10 kGy (2). In order for the technique to be useful in a practical situation, itl is important that the radiation-induced signal is stable throughout the shelf-life of the food. Altough there is somme slight decrease in signal intensity after storage for 28 days at +5°C, the signal is still easily detected (Table 2). When irradiated (2.5 an 5.0 kGy) chicken is cooked, the intensity of the signal in the tibia bones is reduced by about 20 % but is still easily measured (3).

Table 1: Effect of irradiation dose on the ESR signal strength from irradiated chicken bones (tibia).

(Values are the mean of 50 observations)

	Dose (kGy)					
	2.5	5.0	7.5	10.0	SEM	SIG
ESR integral	0.87	1.69	2.36	3.09	0.038	***

SEM = standard error of a mean
SIG = significance of effect
*** = P<0.001

Table 2: Effect of storage time and storage temperature on the ESR signal strength from irradiated chicken bones (tibia).

(Values are the mean of 20 observations)

	Storage time (days)						
Temp °C	0	7	14	21	28	SEM	SIG
-20°	2.03	2.08	2.05	1.96	2.01		
+ 5°	2.27	1.87	2.01	1.05	1.92	0.061	**

SEM = standard error of a mean
SIG = significance of effect
*** = P<0.001

The free radicals are trapped in the crystal structure of the bone and the age of the chicken at the time of slaughter may influence the reponse to irradiation. As the bird ages bone crystallinity would be expected to increase. This has been confirmed with irradiated (2.5, 5.0, 7.5 and 10 kGy) tibia bones from birds of 4, 5, 6, 7 and 8 weeks of age (4). Using crystallinity ratio (5) as a measure of bone crystallinity it has been shown that with increasing age, the bones become more standard crystalline (Table 3). Calculation of the ESR integrals to a standard crystallinity coefficient reduced but did not totally account for all the variation due to age.

Table 3: Effect of age on the ESR signal strength and crystallinity coefficient of irradiated chicken bones (tibia).

(Values are the mean of 16 observations)

	Age (weeks)						
	4	5	6	7	8	SEM	SIG
ESR Integral	1.91	2.02	2.10	2.32	2.49	0.064	***
Crystallinity coefficient	2.32	2.42	2.48	2.55	2.60	0.080	**

SEM = standard error of a mean
SIG = significance of effect
*** = $P<0.001$
** = $P<0.01$

In practical sitatuation, it is possible that other bones within the carcass could be used for detection of irradiation. Tibia and femur bones irradiated at 5 kGy gave radiation-induced signals of similar intensity while those associated with the ribs and sternum were significantly lower (Table 4).

The technique has also been used successfully to differentiate between two blocks of frozen mechanically deboned turkey meat, one of which was irradiated using a linear accelerator while the other was not irradiated (6). Extraction of the bone fragments from the irradiated sample gave a signal similar to that shown in Figure 1.

In conclusion, the technique is applicable for the qualitative detection of irradiated food containing bone and research to date has indicated thas a number of factors may influence the intensity of the radiation-induced ESR signal.

Table 4: Effect of bone site on the ESR signal strength from irradiated (5 kGy) chicken bones.

(Values are the mean of 72 observations)

	Bone site					
	Tibia	Femur	Ribs	Sternum	SEM	SIG
ESR Integral	1.80	1.63	1.01	0.85	0.120	***

SEM = standard error of a mean
SIG = significance of effect
*** = P<0.001

REFERENCES

1 - STENVENSON M.H. and GRAY R (1989). An investigation into the effect of sample preparation methods on the resulting ESR signal from irradiated chicken bone. J. Sci. Fd. Agric., 48, 261-267.

2 - STEVENSON M.H. and GRAY R (1989). The effect of irradiation dose, storage time and temperature on the ESR signal in irradiated chicken drumsticks. J. Sci. Fd. Agric., 48, 269-274.

3 - GRAY R. and STEVENSON M.H. (1989). The effect of post-irradiation cooking on the ESR signal in irradiated chicken drumsticks. Int. J. Fd. Sci. Tech.

4 - GRAY R., STEVENSON M.H. and KILPATRICK D.J. (1990). The effect of irradiation dose and age of bird on the ESR signal in irradiated chicken drumsticks. Radiat. Phys. Chem., in press

5 - OSTROSWSKI K., DZIEDZIC-GOCLAWSKA A., STACHOWICZ W. and MICHALIK J. (1974). Accuracy, sensivity and specifcity of electron spin resonance analyses of mineral constituents of irradiated tissues. Ann. N.Y. Acad. Sci., 238, 186-201

6 - gray R. and STEVENSON M.H. (1989). Detection of irradiated deboned turkey meat using electron spin resonance spectroscopy. Radiat. Phys. Chem. 34, 899-909

Paramagnetic Species Induced in Bioapatites by Foodstuff Ionisation

A.M. ROSSI *#, G. POUPEAU*, O. CHAIX*, J. RAFFI˙, J.P. AGNEL˙ and A. JEUNET[º]

* U.R.A. 69, C.N.R.S. and Institut Dolomieu, Université Joseph Fourier, 15 rue Maurice Gignoux, 38031 Grenoble, France.

˙ Département de Physiologie Végétale, C.E.N. de Cadarache, 13108 Saint-Paul-lez-Durance, France.

º LEDSS, Université Joseph Fourier, Grenoble, France.

#*Present adress*: Centro Brasileiro de Pesquisas Fisicas, rua Xavier Sigaud 150, Rio de Janeiro, BRAZIL.

INTRODUCTION

It is known since more than thirty years that ionising radiations induce a signal in the Electron Spin Resonance spectrum of bones (Gordy et al. 1955). The stability of this signal at ambiant temperatures allowed applications to calcified tissue research (Ostrowski et al. 1980), nuclear bomb and accident dosimetry (Ikeya et al. 1984, Ikeya and Ishii 1989), geological dating (Ikeya 1988, Poupeau and Rossi 1989) as well as the detection and dosimetry of ionised foodstuffs (Dodd et al. 1985).

Radiation dosimetry of calcified tissues relies on the stable signal produced in their mineral (hydroxyapatite) part. However, in spite of extensive and successful applications since many years, the very nature of the ESR spectrum of irradiated bioapatites is not yet fully elucidated. It was suggested that the ESR spectrum of bones and teeth enamel be attributable to one paramagnetic species presenting an axial symmetry (Ostrowsky et al. 1980), an interpretation still recently held by Kai et al. (1988) and Shimano et al. (1989). Alternatively, it was proposed that the asymmetric signal of bioapatites be due to one species characterized by an orthorhombic symmetry (Bacquet et al. 1981). More complex models infer the contribution of several species: either one isotrope species and another one with an axial symmetry (Doi et al. 1979), or at least two species having an orthorhombic symmetry (Callens et al. 1987).

B. Catoire (Editor)
Electron Spin Resonance (ESR) Applications
in Organic and Bioorganic Materials

Working on ***single fragments*** of marrow-free irradiated frog's leg bones, Raffi et al. (1989) observed in X-band ESR spectroscopy a signal in the 2.00 region of the spectrum presenting the characteristiscs of a species having an axial symmetry. Their observations also suggest that in these bones, the hydroxyapatite crystals are preferentially oriented with their crystallographic symmetry axis of order 6 about parallel to the bone elongation direction. Based on X-band (9.5 Ghz) and Q-band (35 Ghz) ESR spectroscopy of single enamel fragments in fossil mammals, Rossi and Poupeau (1989, 1990) showed that the radiation-induced ESR signal in enamel hydroxyapatite is in fact composed of overlapping lines due to two paramagnetic species presenting respectively an axial and an orthorhombic symmetry.

The purpose of this article is to present some comparative observations about ESR spectra of irradiated bones and enamel and discuss the present status of understanding of the ESR spectrum of irradiated bioapatites and its relationships to ionized foodstuff dosimetry.

SIMILARITY OF IRRADIATED FOODSTUFF BONES AND FOSSIL ENAMEL ESR SPECTRA

Unirradiated calcified tissues from vertebrate animals do in general exhibit a flat ESR spectrum. However, unirradiated fresh bones exhibit frequently, when marrow is present, a small signal at g = 2.0042, presenting a peak-to-peak linewidth of 5.8 G (Raffi et al. 1989). Upon gamma-ray irradiation, the intensity of this line increases, while an asymmetric signal, partially overlapping with it, appears on the higher field side of the spectrum (Raffi and Agnel 1989 and this issue, Desrosiers et al. 1990). The latter, with g values of about 2.003 and 1.997, is associated to paramagnetic species developped in the mineral part (hydroxyapatites) of calcified tissues (see e.g. Ostrowski et al. 1974, Stachowicz et al., 1970, Houben, 1971). It is possible, using a specific chemical treatment, to eliminate marrow, which results, in irradiated bones, in the only disappearance of the 2.0042 line, leaving unaffected the two others. The bone ESR spectra presented in this work have been obtained on de-marrowed samples, in order to study the characteristics of the bioapatites-related species. The chemical (de-marrowing) treatment we used in this work is an attack by hydrogen peroxyde.

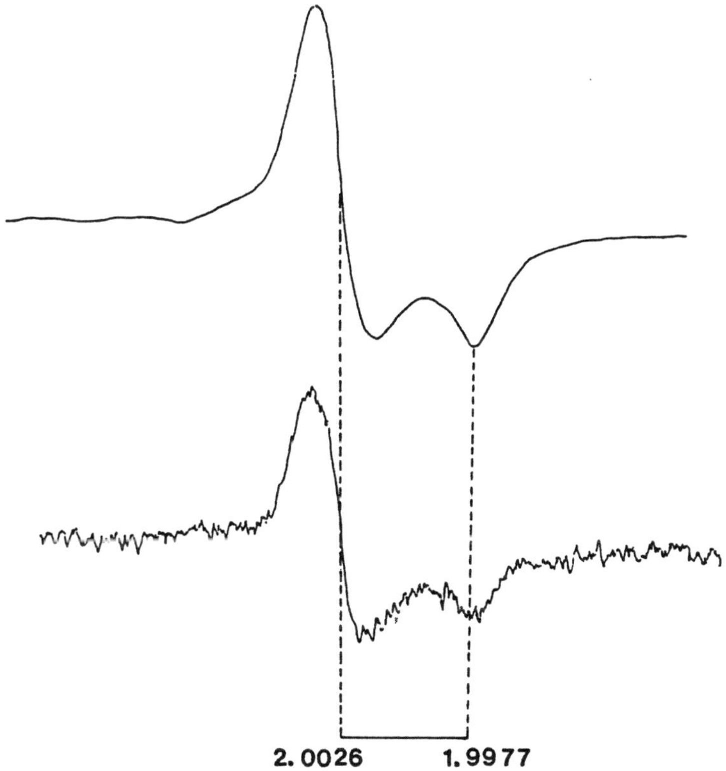

Fig.1 :**X-band ESR powder spectra** of bioapatites. **Top**: gamma-irradiated de-marrowed chicken bone; **bottom**: natural spectrum of an about 1 million years old cervus enamel tooth. The assymetric line presenty g values of 2.0026 and 1.9977.

Irradiated bioapatites show a remarkably reproducible ESR spectrum shape, irrespective of the vertebrate animal species considered. When observed in X-band, the mineral part of fish-bones and marrow-free bones from ionized foodstuffs exhibits an asymmetric signal (Fig. 1, top) characterized by g values of respectively g_1 = 2.0030-2.0033, and g_2 = 1.9969-1.9975 (Raffi and Agnel, this issue).

The same signal is observed in powdered fossil bones and teeth enamel (Fig. 1, bottom), where g values determinations are respectively in the range from 2.0018 to 2.0026 for g_1 and of 1.9977 for g_2 (Yokoyama et al. 1981, Hennig and Grün 1983, Kai et al. 1988, Zymela et al. 1988, Rossi and Poupeau 1989, 1990). It is the sensitivity of this signal to applied radiation doses which is used in the detection of ionized foodstuffs (Raffi et al. 1989 and this issue, Desrosiers et al. 1990, Stevenson and Gray 1989 and this issue) and ESR dating (Ikeya 1988, Poupeau and Rossi 1989). The ESR spectum of fossil enamel is more complex than that of laboratory gamma-irradiated fresh calcified tissues however, as it may also include an organic hyperfine septet characterized by a g value of 2.0033 (Bouchez et al. 1988) and an hyperfine split value of 22 G. This septet, most probably related to an organic radical of the form R-$(CH_3)_2$ (Roufosse et al. 1976), is not induced by laboratory gamma irradiation, nor its intensity -when it is present-, enhanced. Its appearance is therefore attributed to fossilization processes.

Observation of single bones fragments in Q-band reveals that the assymetric signal of bioapatites is complex and results from the superposition of several lines. Again, as in X-band spectrometry, the ESR spectra of powdered bones and enamel present a similar shape (Fig. 2). It is to be noted that the fossil enamel spectrum includes one additional line ("D" line in Fig. 2, bottom) not observed in fresh bone. It is stationary, whatever the orientation of the sample in the ESR resonance cavity be. It therefore belongs to an isotrope species.

The deciphering of complex powder spectra of crystalline materials is possible only if the symmetry and principal g values of the intervening paramagnetic species are known. This supposes that ESR spectroscopy of single, oriented crystals, is feasible. Bioapatites are submicron-sized, and therefore such an approach is not practicable. However, in tooth enamel, needle-shaped hydroxyapatite crystals are locally parallel to nearly parallel to each other. They are elongated, along their c screw crystallographic axis of order six. Thus, it is possible to obtain millimeter-sized enamel fragments behaving, from an ESR spectroscopic point of view, as quasi-oriented systems. We used this opportunity to analyse the nature of the ESR spectra of fossil enamel. Taking into account the similarity of powder spectra of bones and fossil enamel (Figs. 1 and 2), we infer, as discussed below, that the conclusions we arrive at on fossil enamel do apply to ionized foodstuff bioapatites.

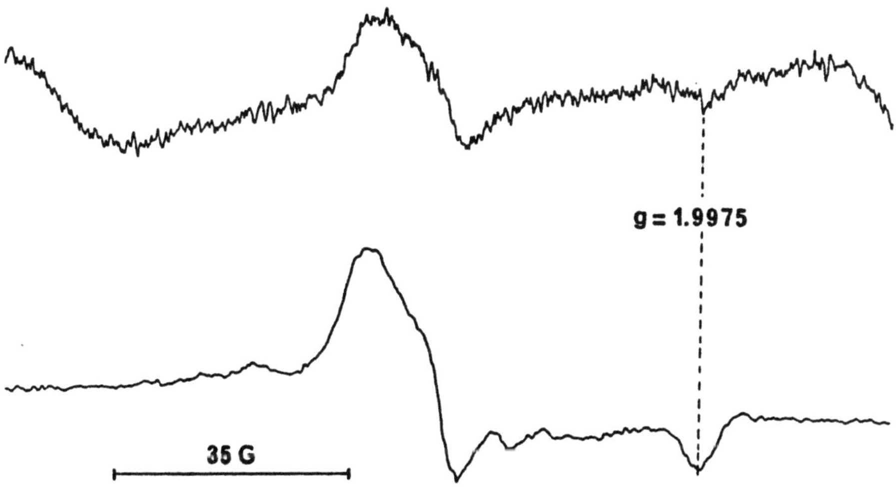

Fig.2 :**Q-band ESR single fragment spectra** of bioapatites. Top: gamma-irradiated de-marrowed frog leg's bone. The bone is oriented with its axis of elongation parallel to H. Bottom: natural spectrum of a 1 million years old Mastodon enamel tooth. The fragment is oriented so that the c axis of the apatites are about parallel to H. In addition to the lines present in Fig. 1, the Q-band spectrum allows to resolve one additionnal line in the enamel. This line, (more visible when c is perpendicular to H, see Rossi and Poupeau 1990) with g = 2.0008, is associated to the radiation-induced isotrope species D of Bouchez et al. (1988).

SINGLE FRAGMENTS SPECTRA ANALYSIS

DE-MARROWED FRESH BONE X-BAND SPECTRA

Raffi et al. (1989) observed that when the spectrometer magnetic field H is rotated in a plane perpendicular to the elongation axis of a frog's leg bone, the X-band ESR spectrum is about conservative. This spectrum shows the same assymetric signal as a powder spectrum. When a section of such a bone is placed with the elongation direction parallel to H and then rotated, the relative intensities of the two components of this line vary with a periodicity of 180°. Raffi et al. conclude that this signal might be related to a species having an axial symmetry. However, the simple fact that in any sample orientation, the two components of this signal are always present strongly suggests that, either hydroxyapatites have various orientations in elongated bones, -although possibly with a preferential direction-, hence a pseudo-powder spectrum, or/and more than one species are present, whose lines might overlap.

Fossil enamel X-band spectra

Rossi and Poupeau (1989, 1990) studied the angular variation of the g factors in single enamel fragments, relatively to the c crystallographic axis of apatites. When the magnetic field of the ESR spectrometer is rotated in a plane containing the c axis of apatites, they observed, in X-band, two apparently stationary lines at g = 2.0026 and g = 1.9977, and one line moving between these two positions, with a periodicity of 180°. When H is rotated in a plane perpendicular to c, the ESR spectrum keeps constant and similar to the bottom spectrum of Fig. 1.

After a thermal annealing in moderate conditions, (e.g. 2 hours at 300°C), angular variation studies show that the "stationary" lines disappeared, leaving only one line, from a paramagnetic species presenting an axial symmetry about the c axis (Fig. 3). The principal g values of this species have been measured in enamel from various fossil mammal species. They are respectively of $g_{\perp} = 2.0026$ and $g_{//} = 1.9977$. According to the crystallographic structure of apatite, Rossi and Poupeau (1989, 1990) suggested that the associated A center, possibly a CO_3^{3-} group, be located at an OH^- site. In addition, one can observe on Fig. 3 a new line, with a g value of 2.0058, corresponding to the isotrope species C of Bouchez et al. (1988). This line, sometimes observable in the natural spectrum of fossil enamel, is always present in thermally annealed samples (Rossi and Poupeau 1990).

Fossil enamel Q-band spectra

Q-band spectra taken on un-heated samples, in the same conditions as in Fig. 3, confirm the existence of one line moving between two extreme positions at g = 1.9975 and about 2.003. In addition, when the c crystallographic axis of apatites are about parallel to H (position $\theta = 0°$ in Fig. 3), it appears clearly that the "stationary line" at g = 2.0026 in X-band is in reality a doublet, "B", as already observed by Bouchez et al. (1988). This is shown in Fig. 4. The two lines of this doublet present g values of respectively $g_1 = 2.0030$ et $g_2 = 2.0018$.

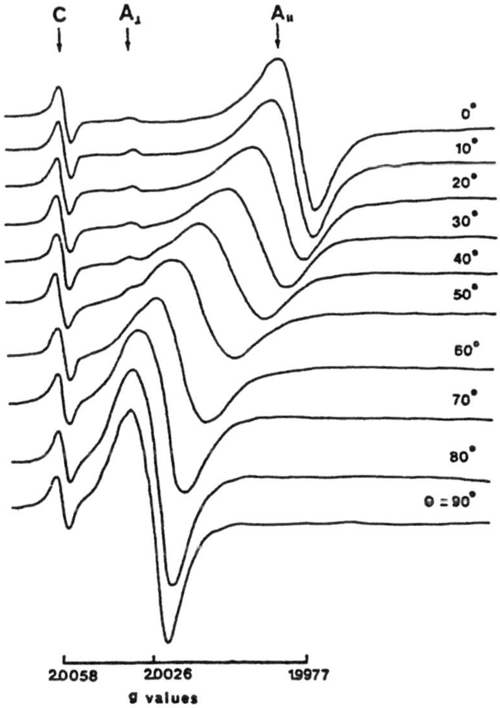

Fig.3: **X-band spectra of one thermally annealed (2 hours at 300°C) fragment of the Mastodon tooth of Fig. 2. The angular variation of the g factors has been studied when the field H of the spectrometer is rotated in a plane containing the crystallographic axis of order 6 of apatites. It reveals the presence of a species, A, presenting an axial symmetry and characterized by g = 2.0026 and g = 1.9977. The isotrope species C, at g = 2.0058, not visible in Fig. 2, appeared as a result of the heating, while the M organic septet (see text) and the D line totally annealed (reproduced from Rossi and Poupeau, 1990).**

In most fossil enamel samples, this doublet is only a minor component of g = 2.00 assymetric signal. However, we found one sample, a shark tooth, where signal A was quite weak relatively to B. In this sample, presenting exceptionnally well aligned apatite microcrystals prisms, it appeared that the B signal is composed of three lines (Fig. 5). Angular variations studies showed (i) a movement of two lines, B_1 and B_2, relative to each other, with a periodicity of 120° and (ii) a variation of intensity of a third line, B_3, with the same periodicity. This suggests that these three lines belong to a single, orthorhombic species B, with $g_1 = 2.0032$, $g_2 = 2.0018$ and $g_3 = 1.9975$.

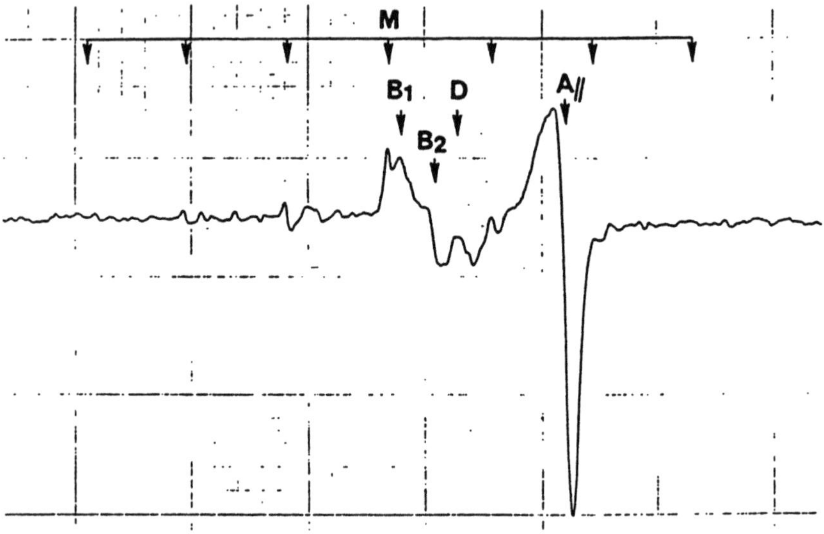

Fig.4 :**Q-band spectrum** of a gamma-irradiated fragment of a Mastodon tooth, observed in the same θ = 0° position as in Fig. 3. Six lines are now apparent, among which those of the isotrope species C and D, the central line of the M organic septet, and in particular the B species doublet at g_1 = 2.0030 and g_2 = 2.0018.

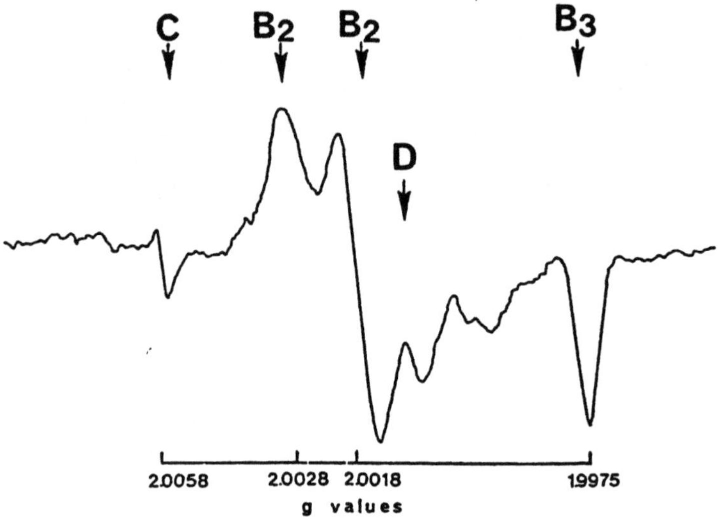

Fig.5 :**Q-band spectrum** of a shark tooth where the B species lines constitute the bulk of the signal. The three lines of the orthorhombic species B, observed in the θ = 0° position of Fig. 3, present g principal values of respectively g_1 = 2.0032, g_2 = 2.0018 and g_3 = 1.9975. (modified, after Rossi and Poupeau, 1990).

The similarity of the g principal values of the orthorhombic species B with that of the axially symmetric species A accounts for the fact that they are not distinguishable in X-band powder spectra, on one hand, and on the other hand, for the co-existency in X-band single fragments, of the superposition of two apparently "stationary" lines and a third one whose position between them depends on the orientation of the c axis of enamel bioapatites relatively to H.

CONCLUSIONS, CONSEQUENCES IN FOODSTUFF ESR DOSIMETRY

NATURE OF THE ESR SIGNAL IN IRRADIATED FOODSTUFF BIOAPATITES

We have shown that strong similarities exist in the g = 2.00 region of the ESR spectrum between powdered bones from gamma-ray irradiated foodstuffs and the natural ESR spectrum of fossil enamel. The X-band spectra of both products are undistinguishable. This similarity is confirmed when the observation in made in Q-band. However, in this case, additionnal lines appear in fossil material: two lines at g = 2.0058 and g = 2.0008, related to the isotropic species C and D, and an organic septet M, at g = 2.0033, with an hf splitting of 22 G (Bouchez et al. 1988). These lines (some of which sensible to radiation doses) were interpreted as consequences of the formation of paramagnetic species related to fossilization processes (Bouchez et al. 1988, Rossi and Poupeau 1989, 1990). X-band and Q-band spectroscopic studies on enamel single fragments allowed to show that, in general, the bulk of the assymetric signal of bioapatites in the g = 2.00 region is attributable to several lines related to two radiation-induced paramagnetic species: species A, with an axial symmetry and g values of respectively $g_{\perp}$ = 2.0026 and $g_{//}$ = 1.9977 and species B, with an orthorhombic symmetry with principal g values of g_1 = 2.0032, g_2 = 2.0018 and g_3 = 1.9975. Considering that the axis of symmetry of species A is parallel to the crystallographic axis of order 6 of apatites, it was proposed that this species be related to a defect (CO_3^{3-}?) located at an OH^- site of the hydroxyapatite lattice ; the orientation of species B in the lattice has not been deciphered and thus its location in the lattice is yet undetermined ; from the similarity of its g principal values with that of species A, it might be possible that these two defects are closely related (Rossi and Poupeau 1990).

In nature, species A and B are directly activated in fossil enamel by the natural radioactivity and the Cosmic Rays. We observed that a laboratory gamma-ray irradiation by a ^{60}Co source results in the increase of their ESR lines, as in the case of foodstuff bone apatites. We therefore consider that the

ESR signal in the 2.00 region of the spectrum is produced, in de-marrowed foodstuff bones and fossil enamel, by the same two paramagnetic species. This conclusion is in agreement with the presence, previously inferred by Raffi et al. (1989), of a species presenting an axial symmetry in irradiated frog leg's bones. It also explains why, in the Raffi et al. observations conditions, a pseudo-powder spectrum was systematically present.

Consequences for ionised foodstuffs ESR dosimetry

The ESR observation of foodstuff bones is not only useful for detecting wether or not they have been irradiated (Dodd et al. 1985), but also to estimate the delivered dose, by successive re-irradiations (Ikeya et al. 1984, Raffi et al. 1989). Despite the fact that the accuracy is sufficient enough to deduce from these measurements the purpose of the initial irradiation, the dosimetry measurement reproducibility is not very good (Desrosiers et al. 1990). Within a single animal, the estimate of the initially applied dose do vary within a factor of 2, according to the bone considered (Desrosiers et al. 1990, Raffi and Stevenson, unpublished results). The differences observed might be, at least partly, due to different contributions of the A, B and organic species to the measured assymetric signal, i.e. to the variations in the relative abundances of mineral and organic constituents of a given calcified tissue. This is especially important if one of the above-related species has a limited lifetime at the temperatures normally used in standard foodstuff storage conditions.

The slight differences observed near to $g = 2.00$ in the X-band ESR spectra of non-demarrowed bones from different animals such as frog, pork and poultry, might also partly result from the same effect. Another important factor in signal analysis, as its intensity is usually taken as the peak-to-peak amplitude of the assymetric apatite line, is the orientation of apatite microcrystals in bones. From these two points of view, frogs elongated bones, where (i) apatites acicular microcrystals seems to be preferentially aligned parallel to the bone axis and (ii) marrow is only present marginally, presently appear as among the most favourable for dosimetric purposes.

Presently, 21 laboratories from the European Economic Community are participating to an intercomparison program on "post-dosimetry" measurements of irradiated foodstuffs, using ESR spectroscopy. This program is sponsored in Cadarache by the Community Bureau of Reference (CBR). A better knowledge of the relative contribution of the A, B and organic species to the structure and intensity of the bioapatite assymetric ESR signal in irradiated calcified tissues would certainly be helpful in the discussions to be held about the variations between "post-dosimetry" measurements. The data presented in this article constitute only a first step in the deciphering of the fine scale study of the ESR signal induced in foodstuff calcified tissues. That also means that new fundamental ESR studies must be carried out, possibly as a future step of the program on identification of irradiated food coordinated by one of us (J.R.) for the CBR.

ACKNOWLEDGEMENTS

Three of us (A.J., G.P. and A.M.R.) are indebted to Drs. J. Lhomme and P. Rey for generous time allocation on ESR spectrometers, respectively at Université Joseph Fourier, Service Chimie/Recherche (LEDSS) and Centre d'Etudes Nucléaires de Grenoble. A.M. Rossi was supported by a post-doctoral fellowship of the brazilian CNPq. Two of us (J.R. and J.P.A.) are indebted to the French Ministery of Research and Technology for a grant n°88 G 104. Laboratory research was partially funded by URA 69, CNRS. This is contribution 1036 of URA 69.

REFERENCES

BACQUET G., QUANG TRUNG V., VIGNOLES M., TROMBE J.C. and BONEL G. (1981)

ESR of CO_2^- in X-irradiated tooth enamel and A-type carbonated apatites,

Calcif. Tissue Intern., 33, 105-109.

BOUCHEZ R., COX R., HERVE A., LOPEZ-CARRANZA E., MA J.L., PIBOULE M., POUPEAU G. et REY P. (1987)

Q-band ESR studies of fossil teeth: consequences for ESR dating,

Quatern. Sci. Rev. 7, 497-502.

CALLENS F.J., VERBEECK M.H., MATTHYS P.F.A., MARTENS L.C. and BOESMAN E.R. (1987)

The contribution of CO_2^{3-} and CO_2^- to the ESR spectrum near g = 2 of powdered human tooth enamel,

Calcif. Tissue Intern., 41, 124-129.

DESROSIERS M., McLAUGHLIN W., SHEAEN L., DODD N., LEA J., EVANS J., ROWLANDS C., RAFFI J. and AGNEL J.P. (1990)

Co-trial on identification and estimates of gamma-ray and electron absorbed doses given to meat bones,

J. Food Sci. Technol., under press.

DODD N.J.F., SWALLOW A.J. and LEY F.J. (1985)

A method of testing for irradiation poultry,

Intern. J. Food. Sci., 26, 451.

DOI Y., AOBA T., OKASAKI M., TAKAHASHI J. and MORIWAKI Y. (1979)

Analysis of Paramagnetic centers in X-Ray irradiated enamel, bone and carbonate containing hydroxyapatite by Electron Spin Resonance Spectroscopy,

Calcif. Tissue Intern., 28, 107-112.

GRAY R. and STEVENSON H. (1989)

Detection of irradiated deboned turkey meat using electron spin resonance spectroscopy,

Radiat. Phys. Chem., **34**, 899-902.

GORDY W., ARD W.B. and SHIELDS H. (1955)

Microwave spectroscopy of biological substances, 1. Paramagnetic resonance in X-irradiated amino-acids and proteins,

Proc. Natl. Acad. Sci., **41**, 983-996.

GRÜN R. (1989)

Present status of ESR dating,

Appl. Rad. and Isotopes, **40**, 1045-1056.

HENNIG G.J. and GRÜN R. (1983)

ESR dating in Quaternary geology,

Quatern. Sci. Rev., **2**, 157-238.

HOUBEN J.L. (1971)

Free radicals produced by ionising radiation in bone and its constituents,

Int. J. Radiat. Biol., **20**, 373-389.

IKEYA M. (1988)

Dating and radiation dosimetry with electron spin resonance,

Magn. Reson. Rev., **13**, 91-134.

IKEYA M. and ISHII H. (1989)

Atomic bomb and accident dosimetry with ESR - Natural rocks and human tooth in vivo spectrometer,

Appl. Rad. and Isotopes, 40, 1021-1029.

IKEYA M., MIYAJIMA J. and OKAJIMA S. (1984)

ESR dosimetry for atomic bomb survivors using shell buttons and tooth enamel,

Japan. J. Appl. Phys., 23, L697-L699.

KAI A., MIKI T. and IKEYA M. (1988)

ESR dating of teeth, bones and eggshells excavated at a Paleolithic site of Douara cave, Syria,

Quatern. Sci. Rev., 7, 503-508.

OSTROWSKI K., DZIEDZIC-GOCLAWSKA A., STACHOWICZ W. and MICHALIK J. (1974)

Accuracy, Sensitivity and Specificity of electron spin resonance analysis of mineral constituents of irradiated tissues,

Ann. N. Y. Acad. Sci., 238, 186-201.

OSTROWSKY K., DZIEDZIC-GOCLAWSKA A. and STACHOWICZ W. (1980)

Stable radiation-induced paramagnetic entities in tissue mineral and their use in calcified tissue research,

in Free Radicals in Biology, Ed. W. Prior, Academic Press, New York, vol. IV, 321-.

POUPEAU G. and ROSSI A.M. (1989)

Electron Spin Resonance Dating,

in Nuclear Methods of Dating, Eds. E. Roth and B. Poty, C.E.A., Paris, 275-293.

RAFFI J. and AGNEL J.P. (1989)

Identification par RPE d'aliments ionisés,

Ann. Fals. Exp.Chim., 82, 279-287.

RAFFI J., EVANS J.C., AGNEL J.P., ROWLANDS C.C. and LESGARDS G. (1989)

ESR analysis of irradiated frog legs and fishes,

Appl. Radiat. Isotopes, 40, 1215-1218.

ROSSI A.M. and POUPEAU G. (1989)

On the nature of the assymetric line in the ESR spectrum of irradiated bioapatites,

Bull. Magnetic Resonance, in press.

ROSSI A.M. and POUPEAU G. (1990)

Radiation damage in bioapatites: the ESR spectrum of irradiated dental enamel revisited,

Nuclear Tracks and Rad. Measurements, in press.

ROUFOSSE A., RICHELLE L.J. and GILLIAM O.R. (1976)

Electron spin resonance of organic free radicals in dental enamel and other calcified tissues,

Arch. Oral Biol., 2, 227-232.

SHIMANO T., IWASAKI M., MIYAZAWA C., MIKI T., KAI A. and IKEYA M. (1989)

Human tooth dosimetry for gamma-rays and dental X-rays using ESR,

Appl. Rad. Isotopes, 40, 1035-1039.

STACHOWICZ W., OSTROWSKI K., DZIEDZIC-GOCLAWSKA A.anf KOMENDER A. (1970)

ESR study of bone tissue sterilized by gamma irradiation,

Nukleonika, 15, 131-142.

YOKOYAMA Y., QUAGEBEUR J.P., BIBRON R., LEGER C., NGUYEN H.V. and POUPEAU G. (1981)

Electron Spin Resonance (ESR) dating of bones of the Caune de l'Arago at Tautavel,

in "Datations Absolues et Analyses Isotopiques en Préhistoire. Méthodes et Limites", Eds. H. de Lumley et J. Labeyrie, CNRS, Paris, 457-492.

ZYMELA S., SCWHARCZ H.P., GRUN R., STALKER A. MacS. and CHURCHER C.S. (1988)

ESR dating of Pleistocene fossil teeth from Alberta and Saskatchewan,

Can. J. Earth Sci., 25, 235-245.

Electron Spin Resonance Spectra of Irradiated Textiles: Applications to the Dosimetry of Ionising Radiations

J. BARTHE, V. KAMENOPOULOU*, B. CATOIRE**, F. BERMANN AND G. PORTAL

Commissariat à l'Energie Atomique, IPSN/DPT/SIDR
BP 6 - 92265 FONTENAY AUX ROSES CEDEX - FRANCE -

* Commission à l'Energie Atomique, Grecque
15310 AGHIA PARASKEVI, ATTIKIS - GRECE -
** Institut Textile de France, Ave. Guy de Collongue,
B.P. 60 - 69132 ECULLY CEDEX - FRANCE -
European Community study contract number : B 160020F

ABSTRACT

In order to be able to evaluate the dose received by a person accidentally exposed to gamma radiation, a method based on observing changes induced in the intensity of the electon spin resonance (ESR) spectrum of irradiated fabrics has been developed. About ten different fabrics were studied: the fibers of three of these fabrics exhibited useful dosimetric properties (cotton, polypropylene, quartz based fibers). In order to minimize the effects of fiber anisotropy on the ESR spectra, a system was devised to rotate samples in the measurement cavity during the time in which the spectrum is recorded. The relation between the intensity of the ESR signal and the ionizing radiation dose is linear over a large dose range (up to a few kilograys). However, the existence of an intrinsic signal from the non irradiated fabric makes it difficult to determine low doses. The minimum dose threshold was raised by employing a special sample preparation technique. The effects of temperature on the ESR signal (thermal fading) from irradiated fabrics was also studied. These results should be considered in the general context of the dosimetry of ionizing radiations through the effects of such irradiations on organic materials. The main results obtained in this field and their usefulness are discussed.

B. Catoire (Editor)
Electron Spin Resonance (ESR) Applications
in Organic and Bioorganic Materials

INTRODUCTION

Since the very beginnings of nuclear energy applications, health physicits have been confronted with the problem of distinguishing between the radiation dose absorbed by detector materials and biological tissues. When one is dealing with a well defined radiation, such as a radiotherapy beam, a simple relation can be established between the effects of radiation on living tissues and the detector, irrespective of the relative sensitivities of either material. It is therefore possible to provide an accurate control of the doses received by radiotherapy patients. In contrast, when the irradiation involves unknown mixtures of photons of various energies, or photons, neutrons and/or charged particles, it is no longer possible to define a simple relation between detector response and the effects of the irradiation on humans unless the detector has a similar radiation response to that of Man. The rather vague term "sensitivity" represents the combined effect of several phenomena : energy absorbtion, spatial (linear energy transfer) and temporal (dose rate) distributions and biological parameters... Only energy absorption is considered here: the most important characteristic of a dosemeter is to have a chemical composition resembling as close as possible that of living tissues. It is for this reason that when organic dosemeters became available in the sixties, health physicists engaged in work on dosimetry believed that, at last, the ideal dosemeter had been devised.

Work was undertaken in this field by the CEA, notably in Professor Rassat's laboratory [1] [2]. An amino acid, "alanine", proved itself to be the most promising dosemeter [3] exhibiting linearity over a very wide range of doses (1 to 10^4 Gy), response stability (a practically non existent fading over a period of one year) and having a chemical composition almost perfectly equivalent to biological tissues. Unfortunately, in the early years ESR spectrometers were neither sufficiently sensitive nor sufficiently inexpensive to be used for radiation protection control. It is for this reason that free radical dosimetry disappeared out of view until the beginning of the eighties : two specific types of need are responsible for the regain of interest in this type of dosemeter:

- industrial applications of radiation,
- accident dosimetry involving high doses.

New developments in dosimetry related to the industrial applications of ionizing radiations are based on early work carried out by the CEA on alanine dosimetry. This new phase of work on alanine was stimulated by a residual water irradiation program in Munich. In the framework of this program, D. Regulla employed capsules containing alanine dosemeters to determine the total dose received by liquid flowing between radiation sources. A new type of dosemeter was subsequently developed by Regulla and his co-workers [4] to be used as a transfer dosemeter (dosemeter sent by post) to calibrate industrial irradiators. In France the LMRI markets the same type of service [5]. These dosemeters are used in the framework of the IDAS (International Dose Assurance Service) program sponsored by the IAEA [6]. A different type of dosemeter has been developed by the japanese, polystyrene being used as a binder [7] instead of paraffine as in France. This dosemeter is marketed under the name "Aminogray". Quite a lot of research has also been done in the field of accident dosimetry by determining doses from free radicals produced in objects worn or carried by the victim. A dramatic fatal accident that occurred in 1982 illustrates the need for this type of dosimetry: an employee working in a Norwegian industrial irradiation facility entered the facility to replace a failed conveyor belt. Unknowingly, the radiation sources were in their irradiation positions during his intervention. His personal dosemeter film was saturated. It was nevertheless possible to evaluate the dose received (40 Gy) from tonicardiac pills that the victim had in his pocket [8]. In such cases treating physicians require to know the dose distribution over the body rather than dose received at a point on the body as is the case with personal dosemeters. Such considerations gave us the idea of using working clothes for dosimetry [9] and justified the extensive research program undertaken by the SIDR (Service for Instrumentation and Dosimetry in Radiation Protection) of the CEA sponsored by the CEC. Free radical dosimetry based on ESR measurements has been studied [10] together with other techniques (stimulated exoelectron emission and thermoluminescence) in the framework of this program.

2. EXPERIMENTAL EQUIPMENT

A commercial ESR spectrometer made by BRUKER, the ER 200 model was used for these studies. It operates in the X band (9 to 10 GHz). In order to overcome difficulties due to the anisotropy of the materials studied, samples are rotated in the resonant cavity using a specially developed device controlled by a microcomputer; the following complementary functions are ensured in this way:

- digitalizing of spectra,
- accumulation and subtraction,
- angular servo-control of sample,
- rotation during measurements.

2.1. Sample rotation [11]

The first two functions enable the stochastic background signal to be reduced. The third function allows anisotropy in a plane, orthogonal to the microwave frequency field, to be studied. The last function homogenizes the signal in this plane by an averaging effect. It should however be noted that the "dynamic" mean signal obtained by rotating the sample does not correspond exactly to the "static" mean obtained by averaging the signal obtained at 10 degree intervals over a complete rotation. Figure 1, shows how the signal varies with different speeds of rotation.

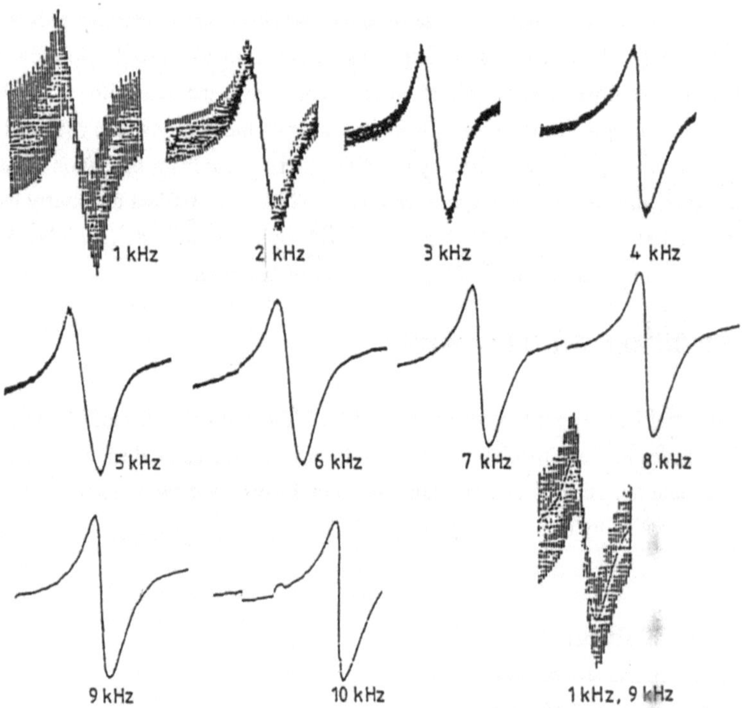

Figure 1 - Effect of sample rotation speed on ESR (3.2 kHz corresponds to 1 rev/second)

The frequencies referred to on the curves correspond to the control frequencies of the step-by-step motor (3200 steps per revolution). The best compromise is obtained at about 10 KHz (3 revolutions per second). If higher or lower speeds of revolution are employed, there is a risk of de-synchronizing the microwave generator servo-system.

2.2. Fabrics studied

The practical objective of these studies is to develop a method to measure the dose received by a person having suffered an accidental irradiation. Emphasis has been given to clothing fabrics. From a historical point of view, cotton fiber based laboratory coat fabrics were the first materials to have been extensively studied. Attention is drawn to the fact that polyesters, which are the principal fiber constituent of much modern clothing, do not exhibit any radiosensitivity.

Polypropylene, wool, linen, thermovyl and even some glass and quartz based fibers have been studied as well as cotton.

2.3. Sample preparation [11]

Two main difficulties became apparent as soon as measurements begun :

- variations in signal strength with the orientation of the sample in the resonant cavity,
- a strong absorbtion of microwave energy by water absorbed by the sample from the surrounding air (function of local relative humidity).

The first of these difficulties is largely overcome by rotating the sample in the cavity (paragraph 2.1.).

The second difficulty is partially overcome by dehydrating samples at 40°C under vacuum for at least one hour. The fabric is then maintained for at least 10 minutes in a dry nitrogen or argon atmosphere at a pressure exceeding that of the atmosphere.

The fabric samples are finally pellitized into small cylinders using a 2 kilobar hydraulic press. In this way, sample isotropy is considerably increased; as the mass of sample in the cavity also increases, the ESR signal is enhanced. Samples in this form are more resistant to external perturbations (air, humidity) because of reduced gas diffusion towards the interior of the sample.

2.4. Irradiation conditions

Most frequently, fabric samples were irradiated before dehydration and compression so as to most closely simulated the conditions encountered in a real irradiation.

Irradiations were performed with a cesium-137 loaded epiirradiator (maximum photon energy : 662 keV, dose rate : 6 Gy per minute).

3. IRRADIATION EFFECTS

The following figures show variations in amplitude and signal shape observed with some of the fabrics studied. Although several different fabrics were studied, only those of cotton [fig. 2] or polypropylene [fig. 3] are sufficiently radiosensitive to be used for dosimetric purposes. Wool [fig. 4], linen [fig. 5] and thermovyl [fig. 6] (non commercial) exhibit intense spectra, which are, however, virtually dose insensitive until doses of the order of several hundred grays are reached; the properties of these fabrics were not therefore studied very extensively.

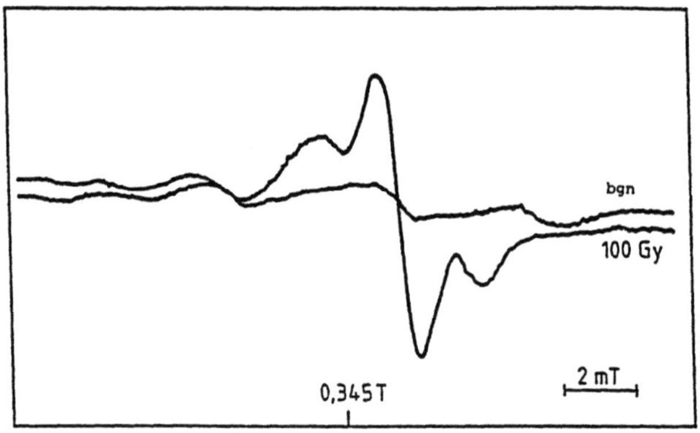

Figure 2 - ESR spectra of ITF cotton (room temperature)

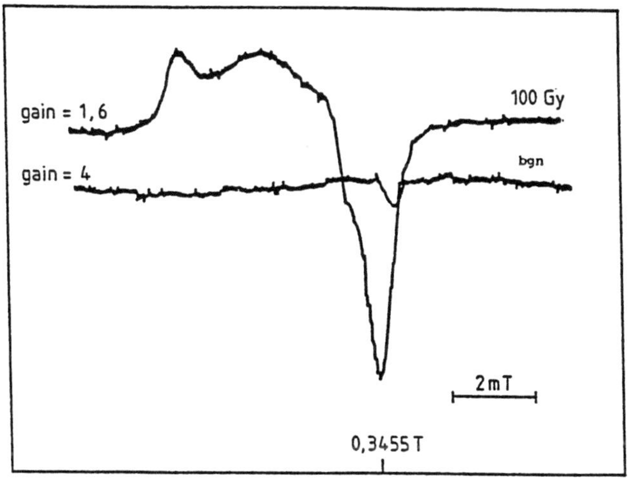

Figure 3 - ESR spectra of ITF PP (room temperature)

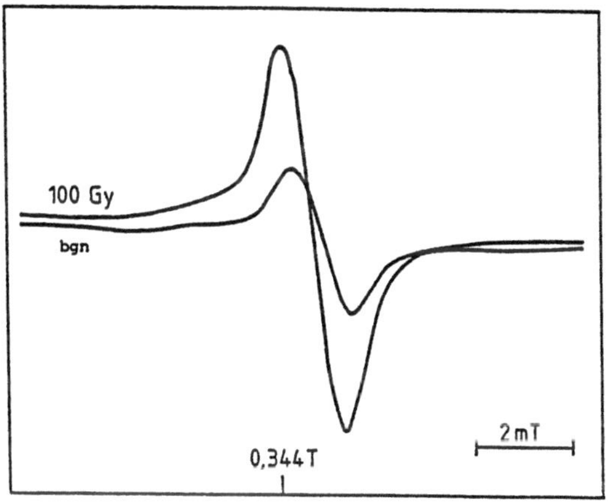

Figure 4 - ESR spectra of wool

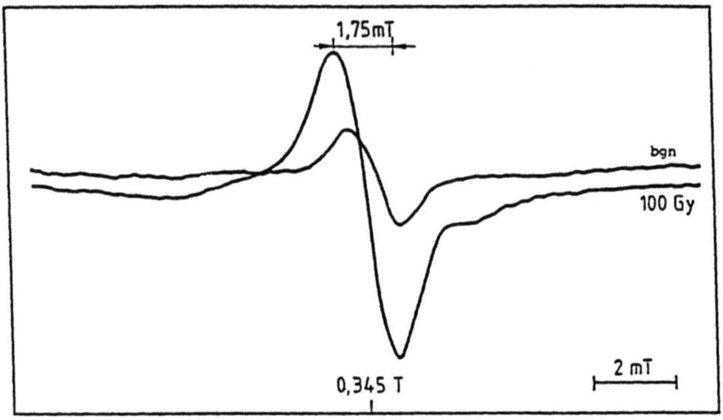

Figure 5 - ESR spectra of linen

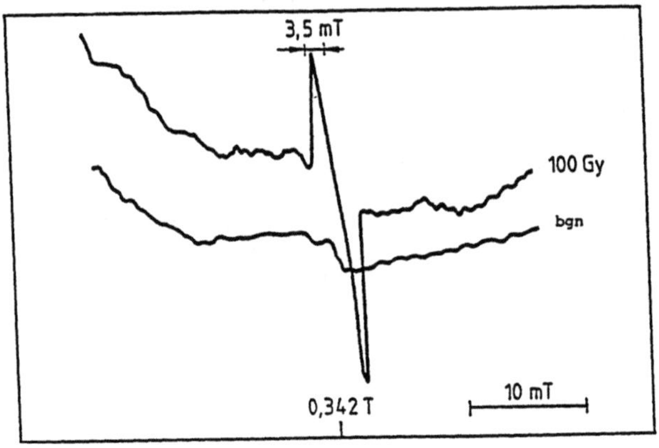

Figure 6 - ESR spectra of thermovyl

3.1. Cotton

Cotton fibers consist mainly of cellulose (92 to 95%), although they also possess a very thin wax-like structure and some peptic matter. The condensed chemical formula of cellulose is :

$$C_6H_{11} - (C_6H_{11}O_4)_n - C_6H_{11}O_5 ,$$

n, being the degree of polymerization. This cellulosic chain is formed by the condensation of β gluclose molecules [12].

Gamma radiation penetrates the glucose and creates secondary electrons of comparatively high energy (a few hundred keV) mainly through the Compton effect ; these electrons lose energy until they become thermalized through interactions with atoms in their environment, excitation, breaking bonds and creating free radicals. A partial de-polymerization of the molecular chains then takes place together with a destruction of the basic cells [13].

After rearrangement free radicals appear. The density of these free radicals gives information about the dose received. Irradiation results in an increase in the amplitude of the corresponding ESR spectrum (fig. 2). However, for convenience and simplicity, ESR intensity measurements reported in the text correspond to the peak-to-peak amplitude of the signal measured and not to some value proportional to the free radical density. As long as the shapes of spectra remain unchanged, the two values are proportional to one another. The ESR amplitude/dose relation curve is linear between a few Grays and 10^4 Gy. Beyond this value the signal amplitude saturates as a result of the radiation destroying as many radicals as it creates [14]. A maximum radical concentration occurs at a dose of 3.10^4 Gy corresponding to a concentration of 10^{17} radicals per cm^3 [15].

The lowest dose detectable corresponds to the intrinsic radical density of the cotton itself and is about 3.10^{13} radicals per cm^3.
Doses are determined by subtracting the response of non irradiated cotton from the signal obtained with irradiated cotton. Under the most favorable conditions a minimum dose of about 1 Gy can be determined.

3.2. Polypropylene (PP)

Polypropylene has a long hydrocarbon chain structure, the condensed formula of which is: $-CH_2 - CH(CH_3-)_n$; PP possesses methyl groups outside its plane. This sterospecific structure enables a variety of different polypropylenes to exist having different spatial arrangements and falling into one of the three following categories of structure: isotactic, syndiotactic or atactic. Isotactic structured PP is mainly used in the textile industry as such a structure confers the fabric with the mechanical and tactile qualities sought. The radicals formed in PP by gamma radiation as well as the shapes of the corresponding spectra depend not only on the temperature during irradiation but also on the readout temperature. The presence of oxygen during irradiation leads to the forming and the transforming of many different radicals; the most dominant radicals involved are those of the peroxide type: the ESR spectra of these radicals is an asymmetric singlet (fig. 3). X-ray diffraction measurements performed on PP both before and after irradiation (up to 1 kGy) were unable to reveal any changes in crystallinity as a result of irradiation [16].

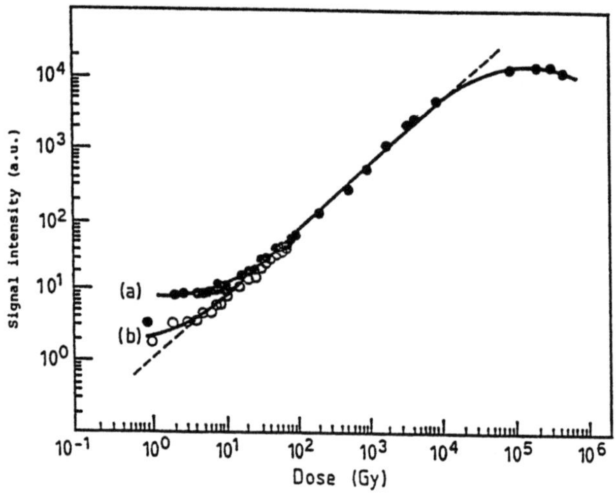

Figure 7 - Signal/dose relation for ITF cotton

• : total signal, o : signal after having subtracted background noise.

The density of radiation induced free radicals (mainly peroxide radicals) gives information about the dose received, as for the case of cotton. The relation between the ESR signal and dose (fig. 8) is linear between a few grays, value beneath which background noise dominates, and saturation, which occurs at about 10^5 Gy.

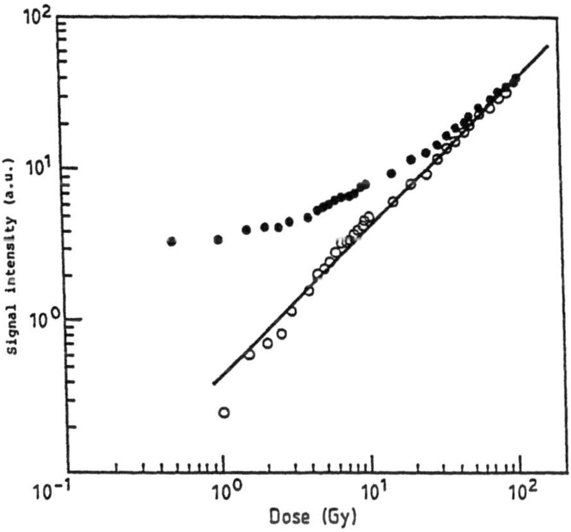

Figure 8 - Signal/dose relation for ICI PP

● : total signal, o : signal after having subtracted background noise.

A non linear relation between signal and dose is observed at low dose values; this is due to a large intrinsic background signal and a low free radical density generated by low doses. However, if the associated background signal is subtracted from the measured signal, the maximum discrepancy from linearity for doses above one gray is about 10%. As with cotton, this limitation is largely due to measurement uncertainties in the background signal resulting from variations in the density of the fabric studied, heterogeneities and textile fiber anisotropies.

4. TEMPERATURE EFFECTS

Generally, the irradiated textiles studied exhibited a slowly, but continuously decaying ESR signal component, termed "fading". In the vast majority of cases this decay in time is less and less significant with decreasing temperature: at liquid nitrogen temperatures the ESR signal remains constant. The decay rate accelerates with increasing temperature. Nevertheless, beyond a certain temperature, the heat generates new radicals leading to an increase rather than a decrease in signal.

The trapped radical concentration decreases with time and finally stabilizes [15] as a result of:

- recombination,
- ions being formed by either capture or loss of electrons,
- transfer of excitation energy to the matrix as a result of the radical dissociating.

4.1. Fading at room temperature

Figure 9 shows how the signal from cotton changes with temperature over a period of 15 days. A rapid decrease in signal is observed during the first ten hours followed by a much smaller or even non existent decrease. A more detailed analysis reveals that the decay of the fast component increases with dose, while that of the slow component is relatively independent of dose.

Similar behavior takes place with polypropylene but the decay is more regular and only one component appears to exist (fig. 10).

4.2. Fading at higher temperatures

These studies have been undertaken in order to investigate whether reversible temperature induced phenomena exist which would enable the initial pre-irradiation signal (background noise) to be determined.

The stepwise recombination model was used to realize this objective. This model is based on the following hypotheses [15,17] :

1. Second order kinetics with the transient formation of a radicular complex.
2. Uniform distribution of radicals as a function of their activation energies.
3. Existence of a capture sphere outside of which there is no recombination.
4. A diffusion velocity that increases with temperature.

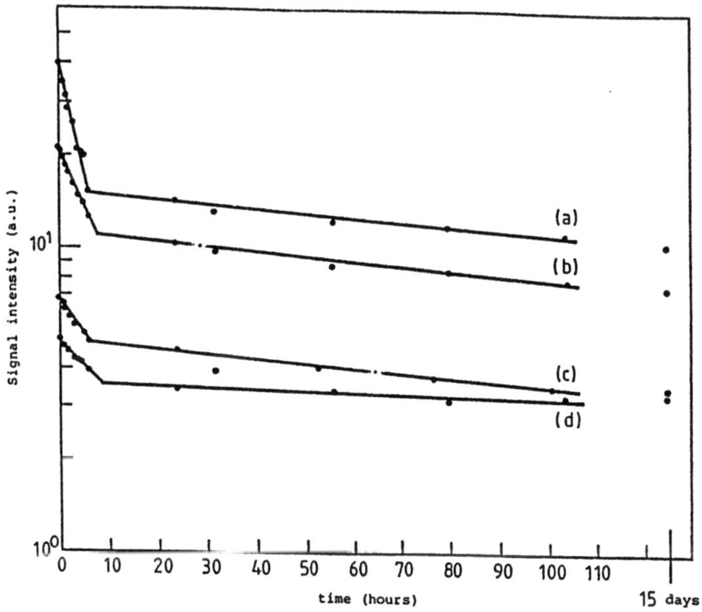

Figure 9 - Cotton fading at room temperature.
(a) 100 Gy, (b) 50 Gy, (c) 10 Gy, (d) 5 Gy

The following conclusions can be drawn :

- the shapes of the recombination curves are, for a given substance, independent of the initial radical concentration or dose received.
- the levels of the "steps" depend on temperature.

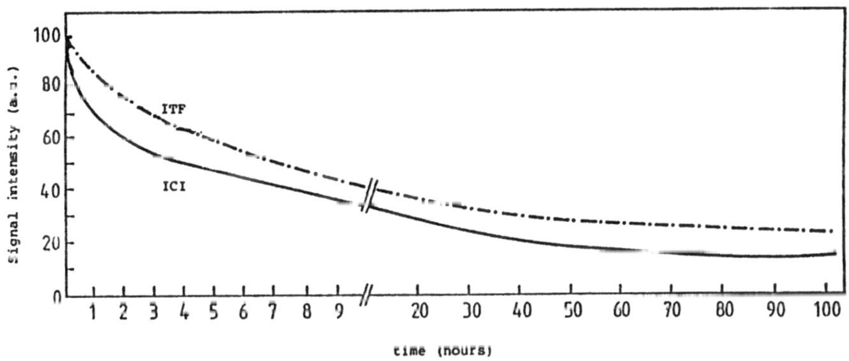

Figure 10 - Fading of various PP's at room temperature.

Figures 11 and 12 show respectively the way in which the ESR spectrum of cotton changes with temperature and how its amplitude varies between room temperature and 205°C over approximately one hour intervals. A decrease in the amplitude of the ESR signal up until about 122°C is first observed; this region corresponds to the thermal recombination of the radicals. Beyond this temperature more and more radicals appear as a result of thermo-chemical reactions.

PP exhibits similar behavior but, in this case, the minimum observed (fig. 13) corresponds to the intrinsic pre-irradiation signal. In addition, it can be seen that the phenomenon is totally reversible and that the irradiation - whitening characteristics remain unchanged as long as the temperature does not exceed 100°C.

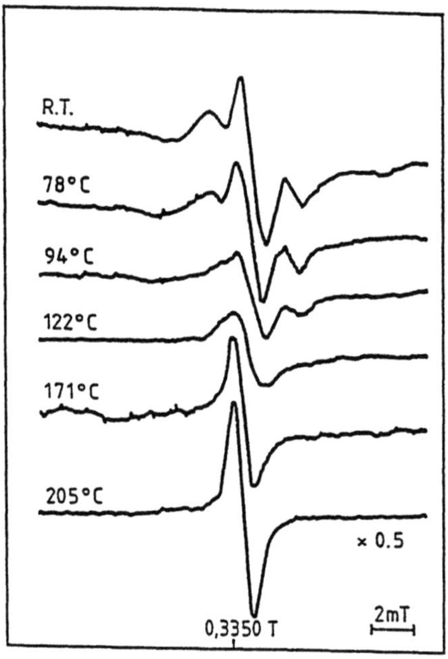

Figure 11 - ESR spectra of cotton as a function of temperature

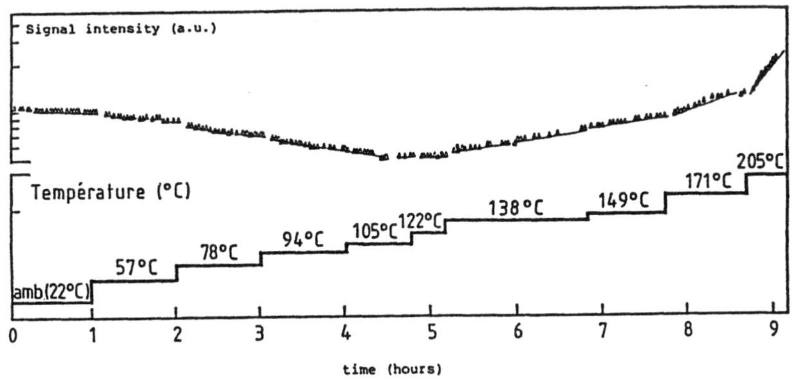

Figure 12 - Effect of temperature on the intensity of ESR spectra of cotton.

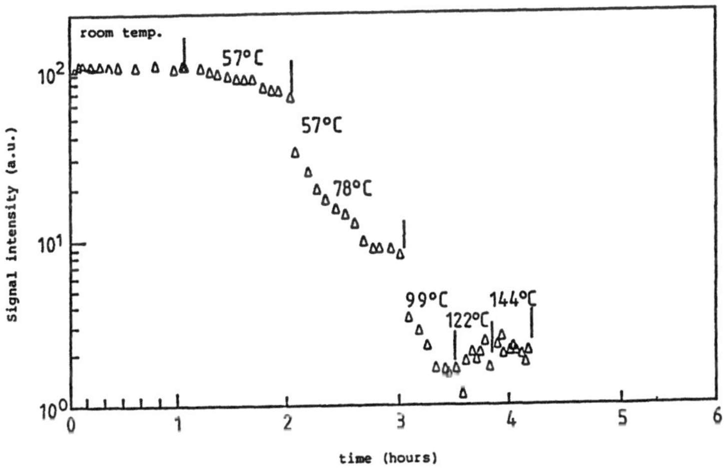

Figure 13 - Effect of temperature on the intensity of ESR spectra of ITF-PP, ICI-PP

Advantage was taken of this effect in order to evaluate the intrinsic pre-irradiation background noise of the fabric. The procedure employed is as follows :

1. Readout of irradiated fabric.
2. Maintaining sample at 90°C for 35 minutes at a reduced pressure.
3. Allowing sample to remain in the laboratory (usual laboratory atmosphere) for 1 hour.
4. Readout of background noise.

Figure 14 shows how the ESR spectrum corresponding to the following conditions :

a) initial background noise,
b) after 100 Gy of irradiation,
c) after heating,
d) after 1 hour in the atmosphere,
e) after heating a second time.

Attention is drawn to the effects of ambient parameters, and in particular, humidity.

4.3. Activation energy

The results that we have just indicated incited us to see whether an Arrhenius modelization based on a Boltzmann distribution could be verified. In this modelization, the intensity of the ESR signal is related to temperature and time by the following relation :

$$S(t) = S_o \,.\, \exp\,[\alpha.(T).t],$$

where S_o is the initial (inaccessible) signal and α the activation constant of the process depending on T.

The ESR signal S(t) practically follows a time dependent exponential decay law. The exponential slope:

$$\alpha(T) = - \Delta t \log S(t)/S_o$$

is determined by applying a least square treatment to the exponential curve for each of the different temperature values.

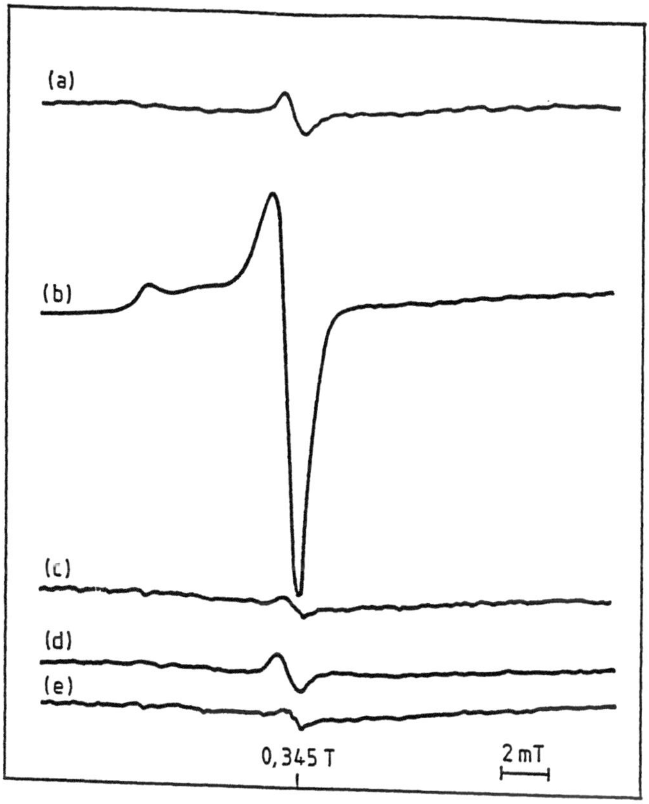

Figure 14.b - Procedure for measuring the initial background noise of irradiated PP

a) initial backgroung noise
b) after irradiation
c) after heating
d) after 1 hour in air
e) after heating for a second time

Applying the Arrhenius model one can now write :

$$\log[\alpha(T)] = \log(\alpha_o) - E/k \times 1/T$$

In this model the slope of the straight line obtained corresponds to E/k where E is the activation energy and k Boltzmann's constant.

The experimental measurements for the two textile fabrics studied are sensibly coherent :

cotton	: E = 0.24 eV
PP(ITF)	: E = 0.63 eV
PP(ICI)	: E = 0.56 eV

5. CONCLUSIONS

This work is a good example fo the applications of ESR to the dosimetry of gamma radiation. Whereas the radicular phenomena occurring when alanine is irradiated are relatively well understood, the situation with organic textile fabrics is quite different. Paradoxically, or may be logically, biological cotton yields results exhibiting less statistical dispersion than the different PP's tested (ITF, ICI, Meraklon, ATO, Marduel, etc...).

In general, fewer thermal processes are generated in cotton than in PP; less thermal fading occurs, and the fading that does occur, tends to disappear after about 10 hours; cotton therefore appears to be more stable. Although the ESR dosimetric characteristics of textile fabrics are not suitable for such materials to be used for personnel monitoring, they are useful for irradiation accident dosimetry in which higher doses are encountered. The obtention of dosimetric information from clothing covering the whole body can provide the physician with invaluable dose distribution information enabling medical treatment to be optimized.

REFERENCES

[1] DESCOURS S., ASSAYRENC J., BERMANN F., COUDERC B., DE CHOUDENS H., DELARD R., RASSAT A., SERVOZ-GAVIN P. Etude par RPE des radicaux libres créés sous irradiation dans certaines substances organiques. Rapport CEA R-3913-1970.

[2] BERMANN F., DE CHOUDENS H., DESCOURS S. Application à la dosimétrie de la mesure par RPE des radicaux libres créés dans les acides aminés. Advances in Physical and Biological Radiation. Detectors AIEA 311-325 - 1971.

[3] KATZ R., BERMANN F. Un dosimètre à large gamme, peu sensible à la qualité du rayonnement : l'alanine. VIIIème Congrès Internat. de la S.F.R.P. - Saclay 1976, 419-434.

[4] REGULLA D., DEFFNER U. Dosimetry by ESR spectroscopy of alanine, Int. J. Appl. Radiat. Isot. 33, 1101-1114, 1982.

[5] SOLLIER T., MOSSE D., CHARTIER M., JOLI J. The LMRI ESR/alanine dosimetry system : description and performance. 2nd Int. Sympos. on ESR Dosimetry - Munich 10-13 Oct. 1988.

[6] NAM J.W. Bulletin de l'AIEA 4 41-43 1988

[7] KOJIMA T., TANAKA R., MORITA Y., SEGUCHI T., Alanine dosemeters using polymers as binders Appl. Radiat. Isot. 37 517-520 1986.

[8] REGULLA D. Assessment of a fatal dose at an industrial Irradiator event. Second International Symposium on ESR dosimetry and applications. Munchen October 10-13 1988.

[9] KAMANOPOULOU B., BARTHE J., HICKMAN C., PORTAL G.
Accidental gamma irradiation dosimetry using clothing.
Radiat. Prot. Dosim. 17 185-188 (1986.

[10] BARTHE J., KAMANOPOULOU V., CATOIRE B. and PORTAL G.
Dose evaluation from textiles fibers. A post determination of initial ESR signal.
1st Int. Conf. on ESR dosimetry, Munich 6-10 october 1988.
A paraître dans Applied Radiation and isotopes.

[11] KAMENOPOULOU V.
Propriétés dosimétriques des fibres textiles : application à la dosimétrie par résonance paramagnétique électronique d'un accident d'irradiation γ.
Thèse d'Université n° 174, Toulouse (Septembre) 1987,
Rapport CEA-R-5425 FRANCE (1988).

[12] OBERLIN M.
Recherches sur la structure et l'organisation cristalline de la cellulose.
Thèse d'état, Paris (1959).

[13] ARTHUR J.C., BLOUIN F.L.A.
Radiation induced graft polymers of cellulose.
In: Proceedings of the International Symposium on Radiation Induced Polymerization and Copolymerization,
Colombus, Nov. 29-30, TID-7643, pp. 319-334 (1962).

[14] DALTON F.L., HOULTON M.R., SYKES J.A.
Gas yields from electron irradiated cotton cellulose.
Nature, 200, pp. 862-864 (1963).

[15] PSHEZHETSKII S.V.
EPR of free radicals in radiation chemistry.
New York : J. Wiley (1974).

[16] SOTTON M., ARNIAUD A.M., RABOURDIN C.
Etude, par diffractométrie des RX, des paramètres de cristallinité et de désordre dans les fibres textiles.
Bull. Scient. ITF, 7 (27) pp. 265-303 (1978).

[17] DILLI S., ERNST I.T., GARNETT J.L.
Radiation induced reactions with cellulose.
Austr. J. Chem., 20, pp. P11-927 (1967).

The Use of EPR Spectroscopy in the Molecular Approach to the Preparation of Supported Catalysts and to Catalytic Processes

Catherine Louis, Michel Che,
Laboratoire de Réactivité de Surface et Structure, URA 1106 CNRS
Université P. et M. Curie, 4 place Jussieu, 75252 - Paris Cedex 05, France

and Zbigniew Sojka
Jagellonian University, ul. Karasia 3, Krakow, Poland

ABSTRACT

This review* shows that EPR spectroscopy can be a powerful tool for catalysis research. This is illustrated in the particular case of highly dispersed Mo/SiO_2 catalysts prepared by grafting. The different steps of the preparation may be identified : for example, the EPR changes observed in the Mo^{5+} coordination sphere during the grafting process of $MoCl_5$ onto silica, showed that the grafted species is probably $SiOMoCl_4$ ($g_\perp = 1.952$, $g_{//} = 1.968$). The catalyst was also characterized after activation treatment such as thermal reduction. Three different Mo^{5+} species have been detected. Their coordination sphere was determined using probe molecules such as water, methanol and natural or ^{13}C enriched carbon monoxide. These Mo^{5+} species are molybdenyl compounds surrounded by oxygen ligands in distorted octahedral (Mo^{5+}_{6c}) ($g_\perp = 1.944$, $g_{//} = 1.892$), square-pyramidal (Mo^{5+}_{5c}) ($g_\perp = 1.957$, $g_{//} = 1.866$), and distorted tetrahedral symmetry (Mo^{5+}_{4c}) ($g_\perp = 1.926$, $g_{//} = 1.755$). The subscripts 6c, 5c and 4c stand for hexa-, penta- and tetracoordinated, respectively. The EPR results obtained on UV-irradiated oxidized Mo/SiO_2 on which methanol was preadsorbed, are also consistent with the variation of selectivity in methyl formate and formaldehyde with Mo dispersion observed during methanol oxidation. They extend the reaction mechanism previously proposed on the basis of kinetic data. EPR revealed the formation of a hydroxymethyl radical $\cdot CH_2OH$. Its deactivating process by an "external" redox process is favoured on highly dispersed Mo catalysts such as grafted Mo/SiO_2. It involves the radical migration on silica where it can react with $SiOCH_3$ groups, giving rise to the formation of methyl formate.

* also published in J. Chem. Soc. Faraday Trans. I, 85, 3939 (1989)

B. Catoire (Editor)
Electron Spin Resonance (ESR) Applications in Organic and Bioorganic Materials

INTRODUCTION

The main steps in the development of a catalytic process involve the control of all the preparation steps of the catalytic system, the characterization of its bulk and surface physico-chemical properties and finally the measurement of its catalytic properties (essentially activity, selectivity and stability). These steps are usually repeated until the best catalytic system is obtained by trial and error.

The present paper will illustrate how the EPR techniques have contributed to improve our understanding, at a molecular level, of these main steps for catalytic reactions occurring on oxide systems.

The use of powder catalysts generates a number of difficulties particularly the frequent presence of several EPR active species (1). In order to determine their magnetic parameters, it is often convenient to employ adapted techniques such as spectra simulation (1), isotopic labelling (1), Q band (2), and third derivative spectra (3).

This paper is a review of the EPR results obtained on silica-supported molybdenum catalysts prepared in our laboratory, by either the classical impregnation method, or the grafting method. This last method involves the reaction of molybdenum pentachloride with the silanol groups of a high surface silica support.

The experiments details and the full description of the results, if already published, can be found in the original references. In the present review, the surface and coordination properties of the grafted molybdenum ions using various probe molecules will be emphasized before presenting the main steps of the oxidation of methanol used as the catalytic test reaction .

Reduced Mo catalysts exhibit EPR signals due to the presence of paramagnetic Mo^{5+} ions ($4d^1$). If the latter are in interaction, the dipolar coupling broadens their signal and prevents their detection. As illustrated below, the Mo^{5+} signal often exhibits some hyperfine lines on the low field side. This hyperfine structure is due to the interaction between the unpaired electron and the nuclear spin (I=5/2) of the ^{95}Mo and ^{97}Mo isotopes which have about the same magnetic moment (natural abundance 15.8% and 9.5%, respectively).

RESULTS AND DISCUSSION

I - Catalyst preparation

The first EPR studies of supported Mo catalysts were performed with samples prepared by impregnation of the support with $(NH_4)_6Mo_7O_{24} . 4H_2O$, calcined at 773-873K, and then reduced (4-8). The alumina support which is of interest from a catalytic point of view, gives spectra (7, 8) whose poor resolution may be attributed for the most part to the superhyperfine splitting due to the interaction of the unpaired electron of Mo^{5+} with ^{27}Al nuclei ($I = 5/2$). Thus, most of the EPR characterization work has been performed with silica which does not lead to this problem.

1. Impregnation method

The reduction by H_2 (100 Torr) at 773K for 20 min of impregnated Mo/SiO_2 catalysts prepared by the incipient wetness method, led to an EPR Mo^{5+} signal at $g_\perp = 1.940$ and $g_{//} = 1.882$, characteristic of an axial crystal field (9, 10, and Table 1). After reduction by CO (100 Torr) at 873K for 3h, another Mo^{5+} species was detected in addition to the species described above (10, and Table 1).

Table 1 : g values of Mo^{5+} EPR signals and characteristics of reduced Mo/SiO_2 catalysts

samples	Mo weight %	reduction	$g_\perp$	$g_{//}$	ref
impregnated	0.17, 2	H_2/773K	1.940	1.882	9,10
"	"	CO/873K	1.961 1.942	1.861 1.891	10
"	1.03	H_2/873K	1.956 1.945	1.859 1.889	12, 14, 17
pseudografted	0.15, 1.56	H_2/773K	1.958 1.941	1.856 1.885	10
grafted	0.18,0.33,1.05	H_2/873K	1.944 1.957 1.926	1.892 1.866 1.755	12, 14, 17

experimental error : $\Delta g_\perp = \pm 0.002$, $\Delta g_{//} = \pm 0.004$

By comparison with the EPR signals of the isopolyanion $Mo_6O_{19}^{3-}$ ($g_\perp = 1.930$, $g_{//} = 1.919$), ^{95}Mo enriched Mo/SiO_2 catalysts and various molybdenyl compounds, Che et al. (9, 11) showed that both species possess a molybdenyl character. They also showed that the Mo^{5+} coordination spheres were composed of oxygen ligands, in distorted octahedral (Mo^{5+}_{6c}) and square-pyramidal symmetry (Mo^{5+}_{5c}) where the subscripts 6c and 5c stand for hexa- and pentacoordinated, respectively. In the case of Mo^{5+}_{5c}, the vacancy in the Mo^{5+} coordination sphere was located in trans position with respect to the "yl" oxygen. The following models were proposed for these two species :

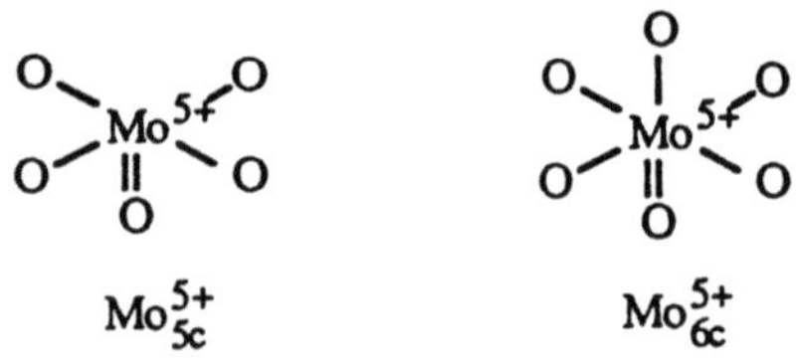

2. Grafting method

This method involves a chemical reaction, under air -and water- free conditions, between surface hydroxyl groups of silica and $MoCl_5$. In contrast to impregnation where the Mo becomes bonded to the support during calcination, the grafting reaction directly produces this bonding which is expected to prevent Mo migration and agregation during calcination and lead to a better dispersed Mo.

In the first study (10), $MoCl_5$ was treated with a suspension of silica in chloroform. The catalyst was then washed with chloroform, further hydrolyzed with water vapour and finally dried at 393K. The samples obtained were blue. After thermal reduction under H_2 or CO, the same Mo^{5+} species were detected as in the case of impregnated samples reduced under CO (Table 1). The blue colour of the samples was due to the presence of molybdenum blues (mixed valence Mo^{5+}-Mo^{6+} hydroxides) in weak interaction with silica, since these hydroxides could be eliminated by water washing. As a matter of fact, the molybdenum was shown not to be really grafted (12, 13), hence the term pseudo-grafted in Table 1.

The grafting method was improved by the use of $MoCl_5$ as vapour (12, 14, 15) or dissolved in cyclohexane (13, 14), a solvent less polar than chloroform. The choice of $MoCl_5$ is particularly attractive since it is monomeric and paramagnetic in the vapour phase and in cyclohexane (16). If not altered during deposition, the Mo^{5+} oxidation state can be used as a very effective probe to monitor, via its EPR spectrum, the changes in the coordination sphere occurring during the grafting process (14). The results presented below concern the preparation in vapour phase. The reactor used for this kind of preparation was equipped with an EPR tube which permitted the monitoring of the grafting process by EPR (12, 14, 15).

When silica (Spherosil XOA 400, Rhône Poulenc, 400m²/g) was heated only in the presence of $MoCl_5$ at 473K, a Mo^{5+} EPR signal with an axial symmetry ($g_\perp$ = 1.952, $g_{//}$ = 1.968, $A_\perp$ = 44 G, $A_{//}$ = 82 G) appeared and increased in intensity (Figure 1). The change of the Mo^{5+} EPR signal before (the monomer $MoCl_5$ gives an isotropic signal at g = 1.952), and after grafting indicates that the Mo^{5+} coordination sphere is affected by the grafting process. The parameters of the EPR signal and of the UV-Vis spectrum of Mo after grafting were found to be very similar to those of the $[MoOCl_4]^-$ ion, suggesting the following grafting reaction (12, 14, 15) :

$$MoCl_5 + SiOH \longrightarrow SiOMoCl_4 + HCl \qquad (I)$$

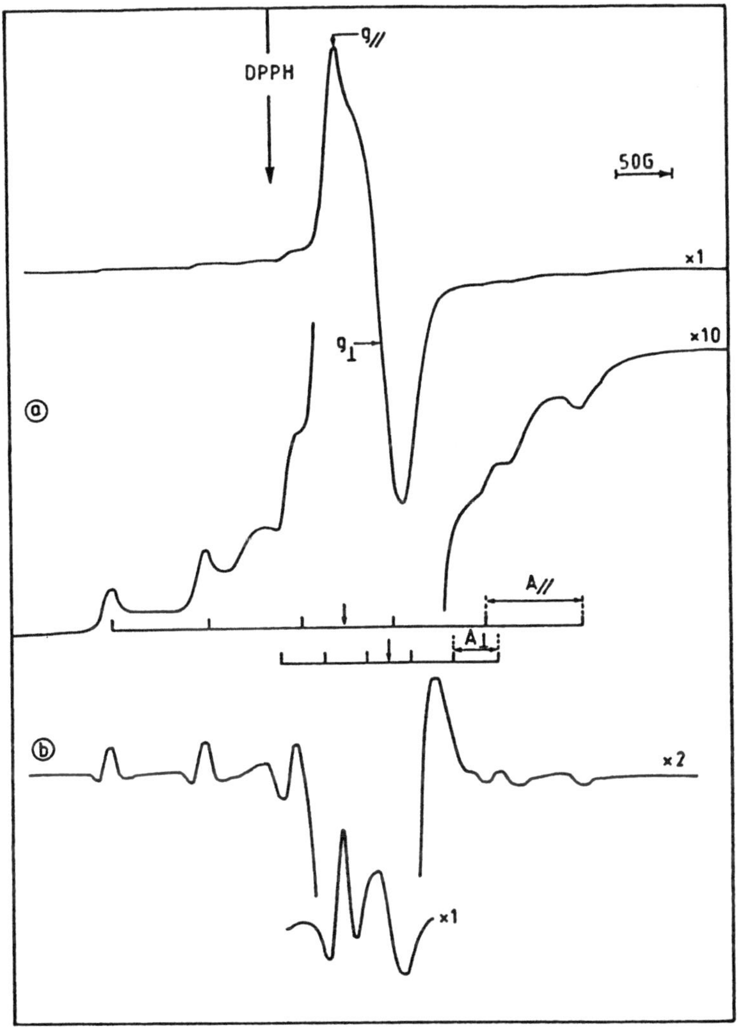

Figure 1 : EPR spectra recorded at 77 K of Mo/SiO_2 catalysts obtained after grafting with $MoCl_5$ vapour at 473K : (a) first-derivative, and (b) third-derivative

The samples turned from red orange after grafting, to a blue colour when exposed to ambient air. This change arose from the hydrolysis and the partial oxidation of loosely bonded $MoCl_5$ into molybdenum blues (12, 15). The latter were eliminated by washing with water or 1M ammonia solutions and the samples turned brown, the colour characteristic of grafted molybdenum. This washing step turned out to be all the more effective that the pH was more basic, increasing the surface negative charge density, thereby removing the negatively charged molybdenum blue species by electronic repulsion (15).

After standard thermal reduction, i.e., in H_2 (200 Torr) at 873K for 2h and further evacuation at 873K for 30 min, the grafted samples exhibited a complex asymmetric line with a mean g value of 1.944 (Figure 2a). The third derivative spectrum (Figure 2b) revealed the presence of three signals (17) whose g values are listed in Table 1. They correspond to three Mo^{5+} species in axial symmetry with $g_\perp > g_{//}$. While two signals were found similar to those of Mo^{5+}_{5c} and Mo^{5+}_{6c} of impregnated samples (Table 1), the third one Mo^{5+}_{4c} was new.

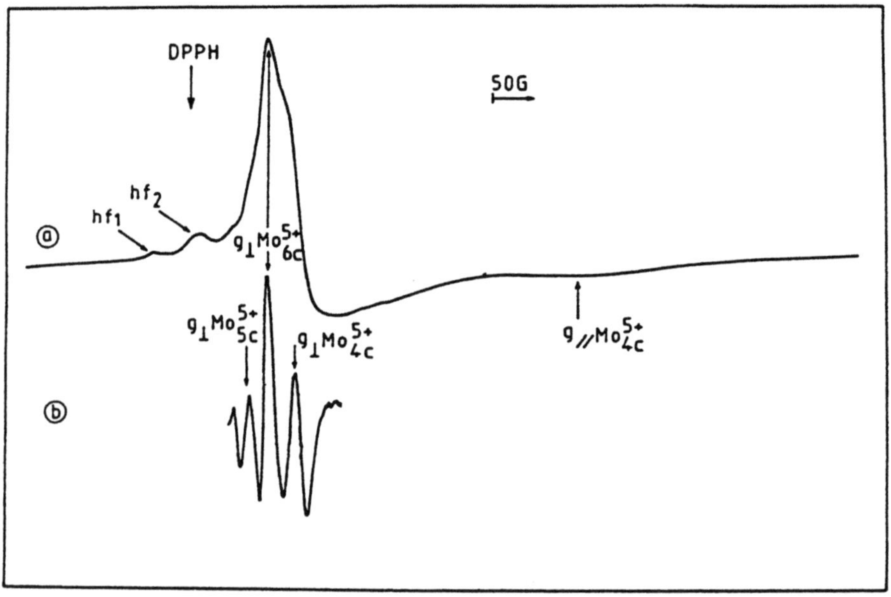

Figure 2 : EPR spectra recorded at 77K of a reduced grafted Mo/SiO_2 catalyst (0.33 wt%) : (a) first-derivative ; (b) third-derivative

II - Molybdenum coordination chemistry

In order to characterize the coordination sphere of the Mo^{5+}_{4c} species and to confirm the symmetry previously attributed to Mo^{5+}_{5c} and Mo^{5+}_{6c}, the approach was to use probe molecules such as water, carbon monoxide (12, 14, 17), and methanol (19) to complete the coordination sphere of coordinatively unsaturated Mo^{5+} ions. As shown below, water and methanol could only provide information on the presence of a coordinatively unsaturated state, whereas ^{12}CO

and ^{13}CO could give information on the number of coordination vacancies through the superhyperfine structure due to the nuclear spin I=1/2 of ^{13}C.

When adsorbed at room temperature, water, carbon monoxide, and methanol only act as ligands and not as redox agents. This was inferred from the constant number of spins before and after adsorption. Methanol is non dissociatively adsorbed on Mo below 350K (19).

1. Water and methanol adsorption

Water adsorption at room temperature led to several transformations in the EPR spectrum recorded at 77K (12, 17) : first, the Mo^{5+}_{4c} signal disappeared whereas that of Mo^{5+}_{5c} increased (Figure 3a), then the Mo^{5+}_{5c} decreased to disappear while that of Mo^{5+}_{6c} increased (Figure 3b). Water desorption at room temperature led to the reappearance of the Mo^{5+}_{5c} signal intensity. At 773K, the water desorption was complete since the three initial Mo^{5+} signals were observed again.

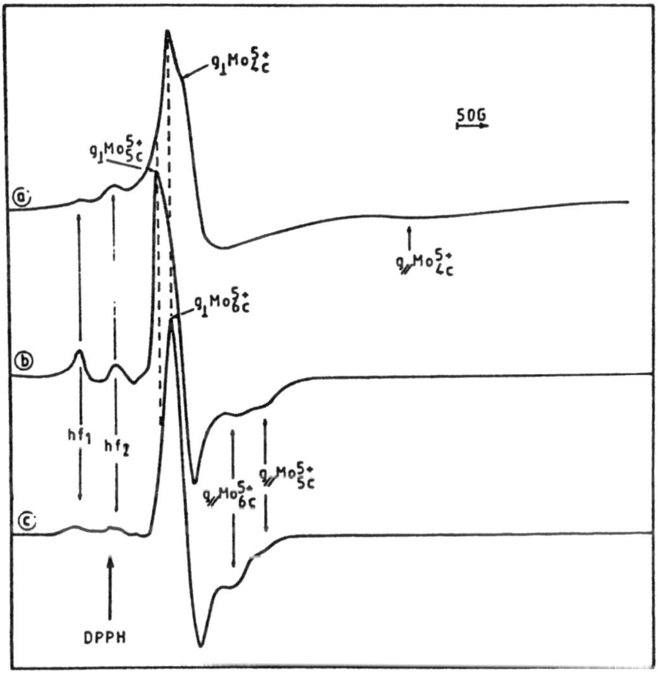

Figure 3 : EPR spectra recorded at 77K of a reduced grafted Mo/SiO_2 catalyst (0.33 wt%) : (a) after reduction ; (b) after water adsorption at 300K (1 Torr) ; (c) after water adsorption at 300K (18 Torr).

Similar results were observed with methanol (18). Variable temperature experiments were performed to monitor the intensity changes of the three signals (Figure 4).

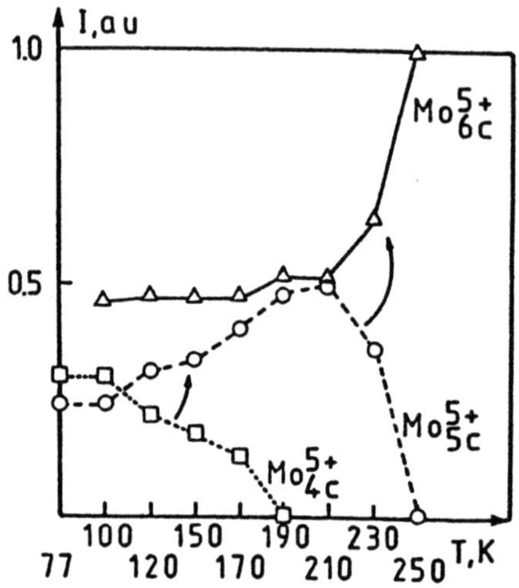

Figure 4 : Temperature dependence of the Mo^{5+} signal intensities in arbitrary units (a.u.), after methanol adsorption on reduced grafted Mo/SiO_2 catalysts (0.33 wt%). The arrows indicate the transformation of less coordinated species to more coordinated ones.

The examination of Figures 3 and 4 shows that H_2O and methanol adsorption is a two-step process (12, 17, 18). The introduction of water or methanol into the Mo^{5+}_{4c} coordination sphere transforms its EPR signal into one analogous to Mo^{5+}_{5c}. Then, the introduction of additional molecules completes the coordination sphere of Mo^{5+}_{5c}, and of the partially coordinated Mo^{5+}_{4c}. This induces the disappearance of their EPR signals and the increase in intensity of a signal similar to Mo^{5+}_{6c} .

These results suggest that Mo^{5+}_{4c} can adsorb more than one molecule of water or methanol and indicate that Mo^{5+}_{4c} contains more than one vacancy in its coordination sphere. The Mo^{5+}_{4c} ion is then the most unsaturated species among the three Mo^{5+} species. Its signal transforming into signals corresponding to molybdenyl ions, the Mo^{5+}_{4c} species is expected to possess also a molybdenyl character.

2. Carbon monoxide adsorption

The counting of coordination vacancies in transition metal ions has been shown to be possible with ^{13}C enriched CO (3, 12, 17, 19) particularly for ground state orbitals like dz^2 or dx^2-y^2 whose lobes are pointing along the bonds with ligands (19).

The adsorption of carbon monoxide (12, 17) induced the disappearance of the Mo^{5+}_{4c} signal and the appearance of a signal composed of two lines at 1.965 and 1.969 whose intensities increased with pressure (until 100 Torr), suggesting that Mo^{5+}_{4c} can also coordinate CO (Figure 5). It was difficult to estimate whether CO could enter in the coordination sphere of Mo^{5+}_{5c} since some of its signal remained visible after CO adsorption. It may be noted that the perpendicular components of Mo^{5+}_{5c} and Mo^{5+}_{6c} signals are less intense than the associated parallel components (Figure 5). This is in contrast to the situation before CO adsorption (Figure 2) or when water was adsorbed (Figure 3). This fact may be explained by the superimposition of the $g_{//}$ (or g_3) components of the new signal(s) induced by CO adsorption either on the $g_\perp$ components of Mo^{5+}_{5c} and Mo^{5+}_{6c} signals, leading to a decrease in their intensity, or on their $g_{//}$ components, inducing in this case an increase of intensity.

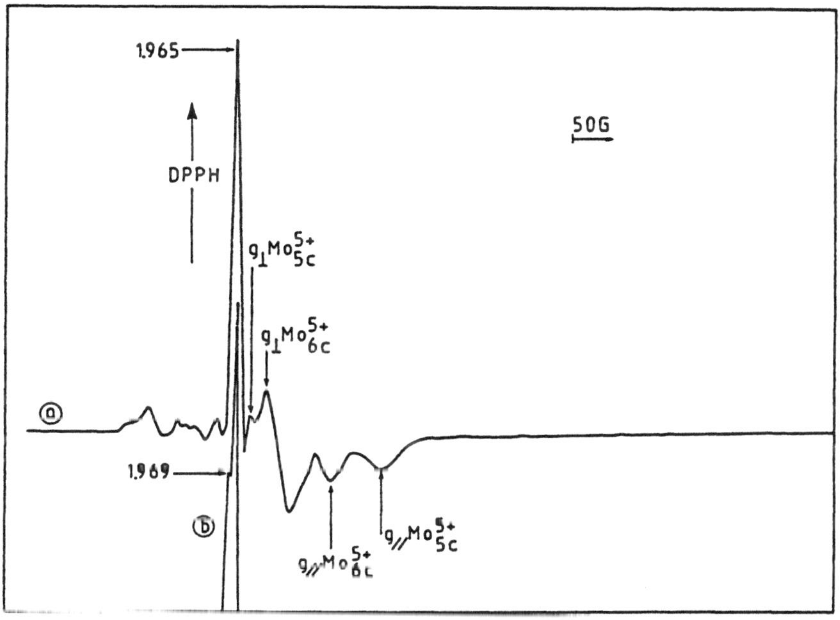

Figure 5 : EPR spectra recorded at 77K of a reduced grafted Mo/SiO_2 catalyst (0.33 wt%) after ^{12}CO adsorption (200 Torr/300 K) : (a) first-derivative; (b) third derivative.

When ^{13}CO was adsorbed, the signal located at 1.965 was broader than with ^{12}CO and the Mo^{5+}_{5c} signal less resolved (Figure 6a). The third-derivative spectrum shows that, instead of two lines, the spectrum is now composed of a quartet, i.e., four lines of different intensities (Figure 6b, 7b), separated from each other by about 7.5 G and located at magnetic fields corresponding to apparent g values of 1.974, 1.969, 1.965 and 1.960. The central two lines of the quartet are located at the same field values as the two lines obtained after ^{12}CO adsorption.

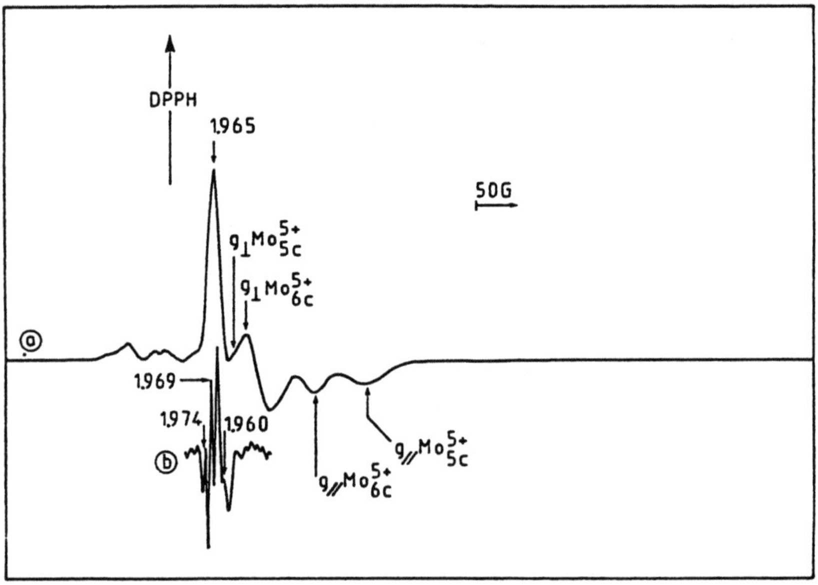

Figure 6 : EPR spectra recorded at 77K of a reduced grafted Mo/SiO_2 catalyst (0.33 wt%) after ^{13}CO adsorption (200 Torr/300 K) ; (a) first-derivative; (b) third derivative.

These four lines were not resolved in Q band (12, 17), suggesting that they arise from a superhyperfine coupling between the unpaired electron of Mo^{5+}_{4c} and the nuclear spin (I = 1/2) of ^{13}C. Indeed, the hyperfine coupling constant is independent of the microwave frequency (2), therefore the splitting observed in X band (7.5 G) is too weak with respect to the line width in Q band (60 G) to be detected.

Assuming similar line widths, these four lines were decomposed into two superimposed triplets centred at 1.965 and 1.969, and with a hyperfine constant of 7.5 G (Figure 7). Using the relative intensities of the two lines obtained after ^{12}CO adsorption, the superposition of the two triplets with relative inner intensities 1:2:1 compares well with the intensities of the experimental spectrum (Figure 7). Each triplet arises from the interaction between the Mo^{5+} unpaired electron and the ^{13}C nuclear spin of two apparently equivalent ^{13}CO molecules. The

Mo^{5+}_{4c} coordination sphere is then completed on admission of two CO molecules. It was therefore deduced that Mo^{5+}_{4c} is a tetracoordinated species :

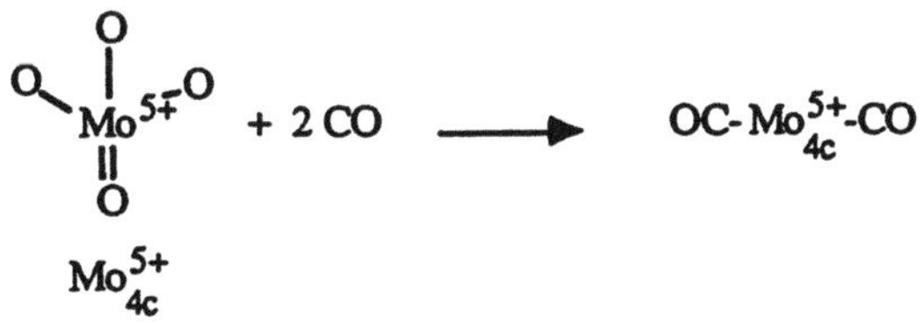

(II)

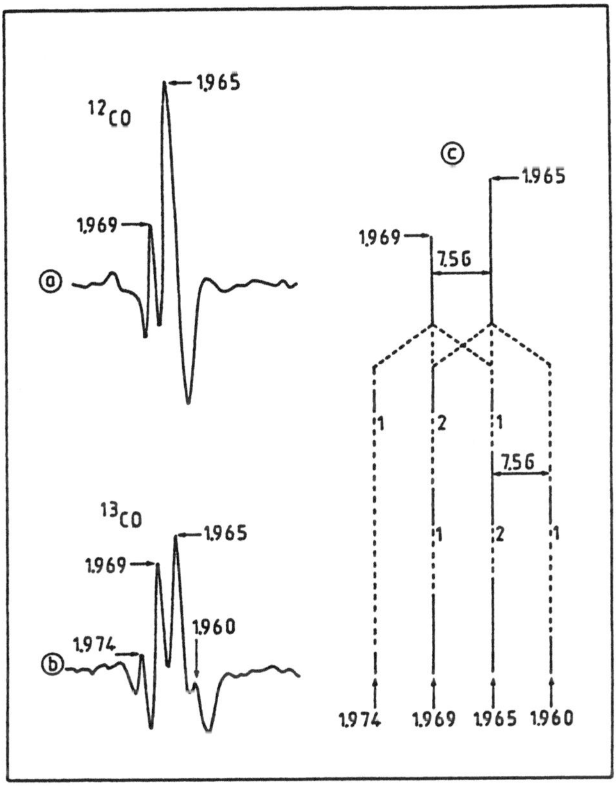

Figure 7 : Third-derivative EPR spectra recorded at 77K of the Mo^{5+}_{4c} ion coordinated by (a) ^{12}CO and (b) ^{13}CO ; (c) is the analysis of these signals.

3. <u>Discussion : the coordination sphere of Mo^{5+} ions after adsorption of gas phase ligands</u>

On the basis of the knowledge of the Mo^{5+}_{4c} coordination sphere, the mechanism of water adsorption can be discussed now. It occurs in two steps :

- the admission of a first water molecule probably located in the equatorial plane since the EPR signal of the partially hydrated Mo^{5+} ion is similar to that of Mo^{5+}_{5c}, which possesses a square-pyramidal structure ;
- the admission of a second water molecule in axial position since the EPR signal of the fully hydrated Mo^{5+}_{4c} is similar to that of Mo^{5+}_{6c} .

The following adsorption mechanism was proposed (12, 17) :

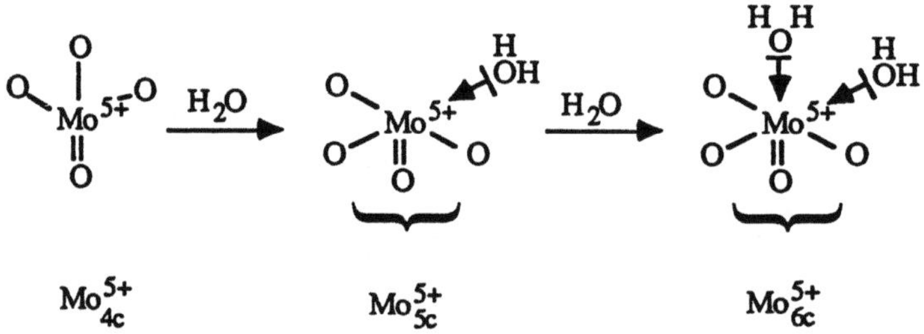

(III)

On the other hand, one water molecule can be adsorbed on Mo^{5+}_{5c} :

(IV)

On the same basis, the following mechanism of methanol coordination by reduced Mo/SiO_2 samples was proposed (18) :

$$Mo^{5+}_{4c} \xrightarrow[120\text{-}190K]{CH_3OH} \underbrace{CH_3\overset{H}{O}\text{-}Mo^{5+}_{4c}}_{Mo^{5+}_{5c}} \xrightarrow[T>190K]{CH_3OH} \underbrace{CH_3\overset{H}{O}\text{-}Mo^{5+}_{4c}\text{-}\overset{H}{O}CH_3}_{Mo^{5+}_{6c}}$$

$$Mo^{5+}_{5c} \xrightarrow[T>190K]{CH_3OH} \underbrace{CH_3\overset{H}{O}\text{-}Mo^{5+}_{5c}}_{Mo^{5+}_{6c}} \qquad \text{(V)}$$

An excess of H_2O or CH_3OH may be adsorbed in the second sphere of Mo^{5+} ions by means of hydrogen bonding or on the silica support by different modes of coordination.

The g values of the three initial Mo^{5+} species and of the hydrated forms can be interpreted on the basis of a $g_\perp$ versus $g_{//}$ diagram (Figure 8) plotted from the g tensor components known for a number of molybdenyl compounds of C_{4v} symmetry (11) : $[MoO\ L_x\ L'_y]^{n-}$, (x=4,5, y=0,1, n=1,2, with L, L' = monodentate ligands). In line with the theoretical expressions of the g tensor components, this diagram (Figure 8) illustrates the dramatic variation of $g_{//}$, mainly due to the spin-orbit coupling constant λ_L of L, in contrast to that observed for $g_\perp$ (11). The following conclusions may be drawn :

- molybdenyl compounds with the same type of ligand L are found in the same region of the diagram, demonstrating the preponderant influence of the λ_L ;
- the molybdenyl halide series expands over both parts of the diagram with λ_L varying from 270 cm^{-1} for fluorine to 5060 cm^{-1} for iodine ;
- the comparison between $MoOCl_4^-$, $MoOCl_5^{2-}$, and $MoOCl_4(H_2O)^-$ shows that the axial ligand has little influence on the g tensor components.
- the oxygen-containing ligands are characterized by $g_\perp > g_{//}$; λ_O, the spin-orbit coupling constant of oxygen, is the same whether the oxygen is an oxide O^{2-} ion or belongs to an hydroxyl group, a water or a methanol molecule. This is the reason why the partially and fully coordinated Mo^{5+}_{4c} species possess EPR signals similar to those of Mo^{5+}_{5c} and Mo^{5+}_{6c}, respectively (Equations III, IV, V).

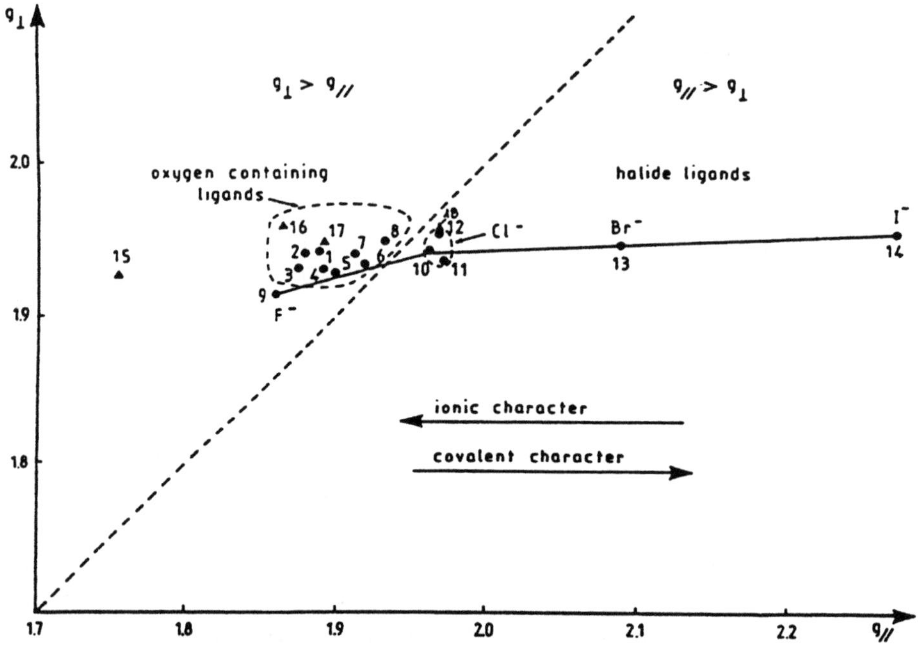

Figure 8 : Representation of various molybdenyl compounds in the $g_{\perp}$, $g_{//}$ plane :
(1) $Mo_6O_{19}H_2$; (2) $MoO(HSO_4)_5$; (3) MoO_3/TeO_3 ; (4) $MoO(H_2PO_4)_5$; (5) $MoO(H_2AsO_4)_5$; (6) $Mo_6O_{19}^{3-}$; (7) $PMoW_{11}$; (8) $MoO(NCS)_5$; (9) $MoOF_5^{2-}$; (10) $MoOCl_5^{2-}$; (11) $MoOCl_4(H_2O)^-$; (12) $MoOCl_4^-$; (13) $MoOBr_5^{2-}$; (14) $MoOI_5^{2-}$; (15) Mo_{4c}^{5+} ; (16) Mo_{5c}^{5+}; (17) Mo_{6c}^{5+} ; (18) $SiOMoCl_4$. Compounds 1-14, (11) and references therein.

By contrast, when a ligand is connected to the central ion by an atom different from oxygen, the spin-orbit coupling constant changes and affects the $g_{//}$ value. This is the case of CO connected to Mo_{4c}^{5+} by the carbon atom (λ_C = 28 cm^{-1} instead of λ_O = 152 cm^{-1}). Moreover, the CO adsorption lowers the Mo symmetry which becomes C_s. The two lines at 1.965 and 1.969 observed after CO adsorption are therefore attributed to the g_1 and g_2 components respectively of an orthorhombic signal ; the g_3 component of the Mo^{5+} carbonyl species, Mo_{4c}^{5+}- $(CO)_2$, as stated above overlaps with the Mo_{5c}^{5+} and Mo_{6c}^{5+} signals and is thus ill-defined. Therefore, the magnetic parameters of the Mo_{4c}^{5+}- $(CO)_2$ are as follows : g_1 = 1.965, g_2 = 1.969, g_3 = ill-defined, A_1 = 7.5G, A_2 = 7.5G, A_3 = ill-defined. The hypothesis of two different Mo^{5+} carbonyl species in axial symmetry may be discarded, since after admission of two CO ligands in its coordination sphere, Mo_{4c}^{5+} no longer preserves its axial symmetry. In addition, the 2 CO appear equivalent (same superhyperfine constant of 7.5 G) and are thus probably located in the equatorial plane. Two models for 2 CO coordinated to Mo^{5+}_{4c} can be proposed :

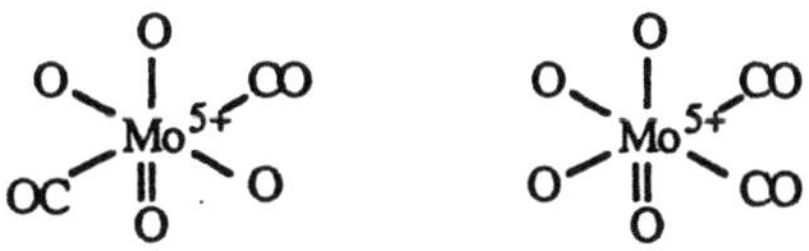

The cis model is more probable since it is formed from Mo^{5+}_{4c} bonded to the surface and that attack by 2 CO is likely to yield the cis isomer for steric reasons.

4. Effect of evacuation temperature directly after grafting

As described above, after grafting in vapour phase, the sample exhibits an EPR signal with $g_{//} > g_{\perp}$ (Figure 1) due to the formation of the $SiOMoCl_4$ species (equation I). If after grafting, the sample was directly evacuated at increasing temperatures without any intermediate exposure to air, the following changes were observed :

- at about 473K, the relative values of the g components are reversed with $g_{\perp} > g_{//}$ (Figure 9a). The spectrum is similar to that of pure Mo^{5+}_{6c} species obtained after water adsorption on reduced sample (Figure 3c);
- at 573K, the spectrum exhibits the signals of both Mo^{5+}_{5c} and Mo^{5+}_{6c} species (Figure 9b);
- at 773K, it becomes similar to that of the reduced samples (Figure 2), i.e., with three Mo^{5+} species (Figure 9c).

The spin concentration measurements indicated that the number of Mo^{5+} ions remains constant during evacuation, suggesting that the spectral changes arise only from modifications within the Mo^{5+} coordination sphere. Figure 8 shows that the inversion of the g values on sample evacuation at 473K, is due to the replacement of the chlorine ligands of $SiOMoCl_4$ by O^{2-} ions of hydroxyl groups or water molecules arising from the silica support, leading to the formation of Mo^{5+}_{6c}. On subsequent increase of temperature, some oxygen ligands are lost, and lead to the formation of vacancies in the Mo^{5+} coordination sphere, i.e., to Mo^{5+}_{5c} and Mo^{5+}_{4c} (12, 14, 15).

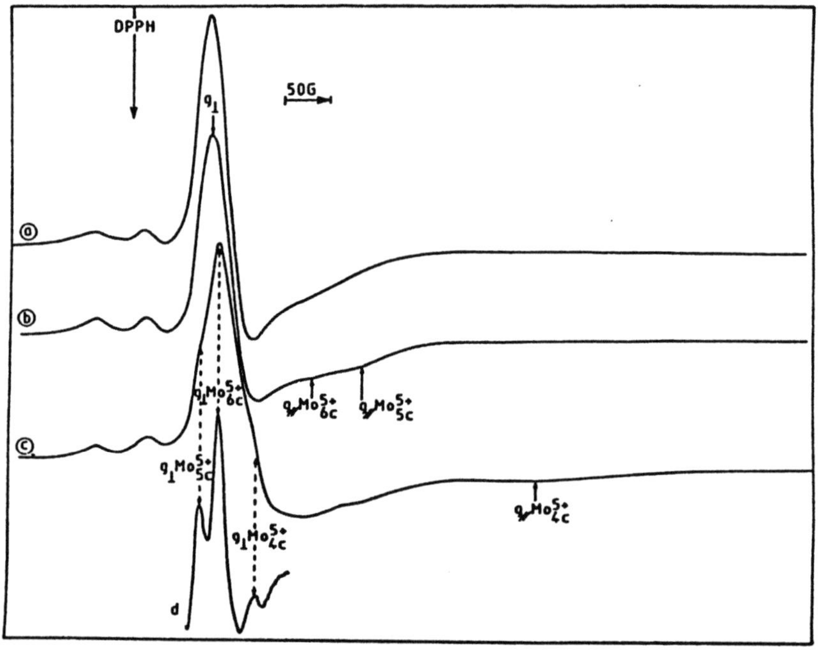

Figure 9 : EPR spectra recorded at 77 K of the Mo/SiO_2 catalysts prepared by grafting with $MoCl_5$ vapour at 473K, followed by evacuation : (a) at 473K for 30 min; (b) at 573K for 30 min ; (c) at 773K for 30 min ; (d) third-derivative spectrum of (c)

Conversely, the Mo^{5+}_{4c} signal disappeared when the sample was left in static vacuum at room temperature during about 24 hours (12, 15) or when water was adsorbed : its signal was first transformed into one similar to Mo^{5+}_{5c} and then into one similar to Mo^{5+}_{6c}. Upon water adsorption, it was shown previously (17) that these changes are due to the admission of a first then a second water molecule within its coordination sphere. The binding with H_2O is via the available doublet of oxygen of the water molecule, the molybdenum remaining paramagnetic. The similarity between the signals of hydrated Mo^{5+}_{4c} species and those of Mo^{5+}_{5c} and Mo^{5+}_{6c} means that Mo^{5+}_{5c} and Mo^{5+}_{6c} ions are also bound to oxygen ligands via dative bonds. These oxygen ligands can belong to OH hydroxyl groups of silica and act as H_2O.

III - Methanol oxidation (18)

Experiments have shown that Mo is much better dispersed on grafted samples than on impregnated samples and in strong interaction with silica (12-15). These characteristics are very attractive to investigate the effect of isolated Mo ions which are believed to be the active sites of certain catalytic reactions. The influence of the Mo relative distance for lower Mo dispersions can also be studied.

Impregnated and grafted Mo/SiO_2 samples have been compared in methanol oxidation (20) which is known to be structure sensitive (21, 22). While on polycrystalline MoO_3, the main product is formaldehyde, on orientated crystallites, the selectivity to formaldehyde, dimethoxymethane or dimethylether depends on the face exposed. In an earlier study (9), the selectivity in formaldehyde and methyl formate, the main products on supported Mo, was shown to depend upon the molybdenum content of grafted catalysts. As the Mo content decreases, i.e., when the Mo dispersion increases, the selectivity in methyl formate increases while that in formaldehyde decreases. In contrast, the dependence is not so clear for impregnated catalysts because of the lack of reproducibility in their preparation. However, their main product is formaldehyde, and this appears to be due to a lower Mo dispersion. The study of the formation of methyl formate using different reaction mixtures and kinetic calculations led us to propose a reaction mechanism. On Mo sites, methanol leads to formaldehyde, which spills over to silica, where it further reacts with methoxy groups to form methyl formate via a hemiacetal intermediate (20).

An EPR study was performed with the aim to establish the elementary steps of this mechanism (18) and to investigate the cycle of model reactions, which may reproduce the most important features of the real catalytic process. As shown above, methanol adsorbed on reduced grafted Mo/SiO_2 enters into the coordination sphere of Mo^{5+}_{4c} (Equation V). No EPR signal appeared when methanol was adsorbed on oxidized sample. However, when the sample was heated above 373K, a Mo^{5+} signal corresponding to Mo^{5+}_{6c} species was observed, indicating that Mo^{6+} ions were reduced.

The molybdenyl bond $Mo^{6+}{=}O^{2-}$ of oxidized Mo is known to be activated as $[Mo^{5+}\text{-}O^{-}]^{*}$ by electron transfer during UV-irradiation at low temperature (77 or 300K) (23, 24) but also by vibrational excitation during thermal treatment (25, 26). The advantage of the low temperature UV-irradiation is to produce stable paramagnetic species, which disappear at higher temperature and are therefore undetectable after thermal treatment.

When the oxidized grafted Mo/SiO_2 samples on which methanol was first adsorbed, was UV-irradiated at 77K, two species were detected by EPR : Mo^{5+}_{6c} ($g_{\perp}$ = 1.946, $g_{//}$ = 1.90) as in the case of thermal activation, and another one as a sharp triplet (g_{av} = 2.003, A_x = 23, A_y = 29 and A_z = 13 Gauss), attributed to the hydroxymethyl radical $\cdot CH_2OH$ (Figure 10). Above 140K, the triplet signal decreased in intensity while that of Mo^{5+}_{6c} increased.

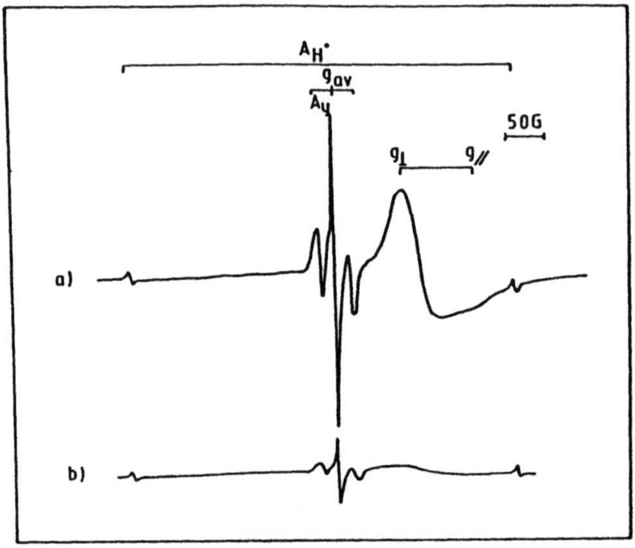

Figure 10 : EPR spectra recorded at 77K, obtained after 10 min of UV-irradiation at 77 K with preadsorbed methanol outgassed to 10^{-4} Torr : (a) reduced grafted Mo/SiO_2 catalyst (0.33 wt%); (b) SiO_2.

The same experiments performed on the pure silica support led to the formation of an EPR signal as a doublet $g_{av} = 2.007$ and A = 140 Gauss corresponding to $\cdot CH_2O$ radicals generated from methoxy groups :

$$SiOCH_3 \xrightarrow{h\nu} SiOCH_2\cdot + H\cdot \qquad \text{(VII)}$$

The EPR results with the Mo/SiO_2 system are consistent with the photoinduced formation of an excited state via ligand to metal charge transfer. This is in agreement with earlier work (23) which have shown that methanol reacts with the $[Mo^{5+}\text{-}O^-]$ * triplet excited state. In our case, methanol was adsorbed first at room temperature and then UV-irradiated at 77K. The following reaction scheme is proposed :

$$Mo^{6+}{=}O^{2-} \xrightarrow[300K]{CH_3OH} \overset{CH_3OH}{\overset{|}{Mo^{6+}}}{=}O^{2-} \xrightarrow[77K]{h\nu} \overset{CH_3OH}{[\overset{|}{Mo^{5+}}\text{-}O^-]^*} \xrightarrow[77K]{h\nu} \overset{\cdot CH_2OH}{\overset{|}{Mo^{5+}}}\text{-}OH^-$$

(VIII)

The last step is a ligand to ligand hydrogen transfer process. It is in accordance with the increase of the Mo^{5+} and $\cdot CH_2OH$ signals upon irradiation and their mutual dependence.

At temperatures above 140K, the $\cdot CH_2OH$ radical becomes unstable. Its decay can be explained in two different ways:

i) deactivation by an "internal nonredox" bi- and/or unimolecular decay of $\cdot CH_2OH$ radicals resulting in EPR invisible species. This process however does not involve any change in the Mo^{5+} ion concentration, in contrast to the experimental results observed.

ii) by an "external" redox decay via interaction with Mo^{6+} ions resulting in their reduction as evidenced by the increase of the Mo^{5+} signal intensity and formation of adsorbed formaldehyde by a ligand to ligand hydrogen transfer process :

$$Mo^{6+} = O^{2-} + \cdot CH_2OH \xrightarrow{T > 140K} CH_2O - Mo^{5+} - OH^- \qquad (IX)$$

This study has shown that the oxidation of methanol is a multi-step process. The order in which these steps occurs, depends on the temperature and the sequence of reactants adsorption, the dispersion and the oxidation state of molybdenum. Other experiments which, for the sake of brevity are not presented here, have confirmed the above processes (18). They involve the adsorption of methanol onto (Mo^{6+}-O^-) formed by N_2O decomposition onto reduced Mo/SiO_2. In the case of thermal activation, the reaction obviously is more complex than for UV-irradiation since methanol can dissociate leading to the reduction of Mo^{6+} into Mo^{5+}.

This EPR study is consistent with the variation of selectivity in methyl formate and formaldehyde with the Mo dispersion observed during methanol oxidation and extends the reaction mechanism proposed on the basis of kinetic data (20). Indeed, on highly dispersed Mo catalysts such as grafted Mo/SiO_2, the decay of the $\cdot CH_2OH$ intermediates by an "external" redox process is favoured. In consequence, these radicals formed at the first stage of the reaction may have enough time to spill over to silica before the final deactivation takes place. This makes it possible for the side reactions with support surface groups, such as $SiOCH_3$, to occur during migration, giving rise to the formation of methyl formate as discussed above. On the other hand, for bulk MoO_3 or low dispersed Mo/SiO_2 catalysts prepared by impregnation, where the Mo adsorption sites are in close interaction and involve at least pairs of reducible Mo^{6+} ions, the two deactivation processes ("internal" and "external redox") are coupled. No migration of $\cdot CH_2OH$ is required for its deactivation to occur and CH_2O becomes the main product of methanol oxidation.

CONCLUSION

The present review has shown that EPR can be a powerful tool:

- during the catalyst preparation, to identify at a molecular level, the different steps of the preparation process. In the case of the grafting of $MoCl_5$ to silica, the changes in the Mo^{5+} coordination sphere have been observed and analyzed;
- to characterize the catalyst after activation treatment such as thermal reduction. Three different Mo^{5+} species have been detected and their coordination spheres determined using probe molecules such as water, methanol and natural or ^{13}C enriched carbon monoxide ;
- to understand the mechanism of the elementary steps occurring during the catalytic reaction, as illustrated here for methanol oxidation. The EPR results have extended the reaction mechanism proposed on the basis of kinetic data.

It is however wise to remember that other techniques can complement the conclusions based on the EPR technique. Owing to its high sensitivity toward paramagnetic species, EPR may at times give a very incomplete, not to say wrong, picture of the main process(es) occurring on catalytic surfaces.

REFERENCES

1 M. Che, and E. Giamello, Spectroscopic Analysis of Heterogeneous Catalysts, J.L.G. Fierro, Ed., Elsevier, Amsterdam, 1989, in press

2 M. Che and Y. Ben Taarit, Adv. Colloid Interf. Sci., 1985, 23, 179

3 M. Che, B. Canosa, and A.R. Gonzalez-Elipe, J. Phys. Chem., 1986, 90, 618

4 G.K. Boreskov, V.A. Dzis'ko, V.M. Emel'yanova, Yu.I. Pecherskaya, and V.B. Kazanskii, Dokl. Akad. Nauk. SSR, 1963, 150, 829

5 J. Masson, and J. Nechtschein, Bull. Soc. Chim. France, 1968, 3934

6 K.S. Seshadri, and L. Petrakis, J. Phys. Chem., 1970, 74, 4102

7 K.S. Seshadri, and L. Petrakis, J. Catal., 1973, 30, 195

8 M. Dufaux, M. Che, and C. Naccache, J. Chim. Phys., 1970, 67, 527

9 M. Che, J.C. McAteer, and A.J. Tench, J. Chem. Soc., Faraday Trans. I, 1978, 4, 2378

10 M. Che. F. Figueras, M. Forissier, J.C. McAteer, M. Perrin, J.L. Portefaix, and H. Praliaud, Proc. Int. Congr. Catal., 6th, 1976, London, 1977, 1, 261

11 M. Che, M. Fournier, and J.P. Launay, J. Chem. Phys., 1979, 71, 1954

12 C. Louis, Doctoral Thesis, Paris 1985

13 C. Louis, M. Che, and F. Bozon-Verduraz, J. Chim. Phys., 1982, 79, 803

14 M. Che, C. Louis, and J.M. Tatibouët, Polyhedron, 1986, 5, 123

15 C. Louis, and M. Che, to be submitted to J. Catal.

16 F.A. Cotton, and G. Wilkinson, Advanced Inorganic Chemistry, Wiley, Ed., New York, 4th Edition, 1980, p.862

17 C. Louis, and M. Che, J. Phys. Chem., 1987, 91, 2876

18 Z. Sojka, and M. Che, J. Phys. Chem., in press

19 L. Bonneviot, D. Olivier, and M. Che, J. Mol. Catal., 1983, 21, 415

20 C. Louis, J.M. Tatibouët, and M. Che, J. Catal., 1988, 109, 354

21 J.M. Tatibouët, and J.E. Germain, J. Catal., 1981, 72, 375

22 J.M. Tatibouët, J.E. Germain, and J.C. Volta, J. Catal., 1983, 82, 240

23 V.B. Kazansky, Proc. Int. Cong. Catal., 6th, 1976, London, 1977, 1, 50

24 M. Anpo, I. Tanahashi, and Y. Kubokawa, J. Chem. Soc. Faraday Trans I, 1982, 78, 2121

25 M. Che and A.J. Tench, Adv. Catal., 1982, 31, 77

26 V.B. Kazansky, Kinet. Katal., 1983, 24, 1338

Application de la RPE à la Charactérisation d'un Matériau Fossile: le Charbon

B.NICKEL-PEPIN-DONAT*(1), A.JEUNET**, A.RASSAT*, H.CHARCOSSET***

*:Ecole Normale Supérieure, 24 Rue Lhomond, 75005 Paris; (1): en détachement au DRFMC / SP2M / PMS, Centre d'Etudes Nucléaires de Grenoble,BP 85X, 38041 Grenoble Cedex.

**: LEDSS, Université J.Fourier, Domaine Universitaire, Batiment 52, BP 58X, 38041 Grenoble Cedex.

*** Institut de Recherche sur la Catalyse, 2, Avenue A.Einstein, 69626 Villeurbanne Cedex.

Le Charbon est un matériau complexe contenant une composante organique et une composante minérale. La RPE peut fournir des renseignements sur ces deux composantes. L'étude des ions paramagnétiques présents dans la phase minérale de certains charbons peut donner des renseignements sur la composante minérale mais la plupart des études portent sur la composante organique dans laquelle la présence de radicaux "fossiles" fut détectée dès 1954 par UEBERSFELD et ses collaborateurs (1).
Nous présentons ici les principaux résultats que nous avons obtenus sur la composante organique des charbons de la mini-banque CERCHAR-GRECO (le Tableau 1 présente quelques caractéristiques de ces charbons). Ces charbons couvrent une gamme de rang allant du Flambant à l'anthracite; ils sont prélevés et conservés sous argon.

Tableau 1

Charbon	C(a)	H(b)	O(c)	S(d)	H/C(e)
Provence	75.25	5.37	13.19	6.72	0.856
Vouters	84.77	5.36	6.7	0.89	0.758
Méricourt	88.7	4.3	3.18	0.82	0.58
Escarpelles	92.5	3.46	2.25	0.82	0.45
La Mure	96.7	1.56	1.93	0.496	0.19

Légende du Tableau 1: les charbons sont classés par ordre de rang croissant (Provence: Flambant, La Mure: anthracite).Pourcentages ramenés à 0% de cendres a): en carbone, b): en hydrogène, c): en oxygène, d): en soufre total; e): rapport atomique hydrogène sur carbone.

B. Catoire (Editor)
Electron Spin Resonance (ESR) Applications in Organic and Bioorganic Materials

I) Etude des charbons bruts

Le spectre de RPE de chacun de ces charbons est constitué d'une raie unique plus ou moins symétrique, de forme généralement ni Lorentzienne ni Gaussienne.Nous avons effectué une étude à deux fréquences (9 et 35 GHz) et une étude en saturation (2). La première étude nous a permis de montrer que tous les charbons présentaient une inhomogénéité en g, la deuxième étude nous a permis de mettre en évidence la présence de deux composantes dans le spectre: une raie fine et une raie large , la composante large ayant une puissance de saturation plus faible que la raie fine.

Le modèle généralement admis pour la partie organique du charbon est un modèle à deux composantes: une composante macromoléculaire (MM) dans les pores de laquelle se trouve une composante moléculaire (M). La composante fine du spectre de RPE est attribuée à la "phase macromoléculaire", la composante large est attribuée à la "phase moléculaire".

Cette étude nous a également permis de montrer que la puissance de saturation (Po) peut être utilisée comme paramètre de rang : elle est corrélable linéairement au rapport H/C et au pouvoir réflecteur de la vitrinite qui sont généralement utilisés comme paramètres de rang

Une étude par RPE de la sorption d'une sonde radicalaire nitroxyde sur ces différents charbons (3) nous a permis de montrer que les charbons pouvaient être classés en deux catégories: les charbons de "bas rang" et les charbons de "haut rang". Les charbons de "bas rang" présentent une phase "fluide" capable de sorber la sonde; la polarité de cette phase "fluide" augmente lorsque le rang du charbon diminue. Une phase "mobile" a été détectée par RMN du 13C, toutefois, il n'est pas possible d'affirmer l'identité de ces deux phases; de la même façon, il n'est pas possible d'affirmer que la composante "large" du spectre de RPE est attribuable à l'une ou à l'autre des composantes "fluide" ou "mobile". Les charbons de "haut rang" présentent, quant à eux, une réactivité vis-à-vis de la sonde radicalaire. Nous attribuons cette réactivité à une réaction de transfert monoélectronique entre la sonde et le charbon.

II) Etude des macéraux des charbons de la mini-banque CHERCHAR-GRECO

Un autre modèle utilisé pour représenter la composante organique du charbon est le modèle "macéral": le charbon est constitué de trois macéraux: l'exinite, la vitrinite et l'inertinite. Tout charbon peut être séparé par centrifugation en gradient de densité en ces trois macéraux (certains charbons ne contiennent que un ou deux de ces trois macéraux). Nous avons étudié chacun des trois macéraux relatifs à chacun des charbons étudiés (4).

Nous avons montré que pour un même charbon, chaque macéral présentait une empreinte RPE propre (facteur g , densité radicalaire, largeur de raie et puissance de saturation). Nous avons également montré que le macéral inertinite des charbons de haut rang présente une grande réactivité vis-à-vis de l'air .
La composante "fine" du spectre de RPE a pu être associée à la présence d'inertinite, la composante "large" étant, quant à elle, associée à la présence de vitrinite et d'exinite. Il est à noter que les deux modèles (macéral et polymère) ne sont pas necessairement en contradiction; en effet, on peut penser que chaque macéral est constitué de chacune des composantes M et MM dans des proportions différentes ou réciproquement. Le modèle "macéral" présente l'avantage de reflèter la constitution du charbon de façon quantitative, ce qui n'est pas le cas du modèle à deux composantes puisque la quantité de composante moléculaire dépend du solvant utilisé pour l'extraire. Le modèle macéral nous a donc servi pour l'interprétation de données obtenues après des opérations d'hydroliquéfaction.

III) Etude de la réactivité d'un charbon en hydroliquéfaction par des observations ex-situ

Le charbon utilisé pour cette étude est le charbon de Freyming qui est équivalent au charbon de Vouters mais qui a été conservé sur parc.
Nous avons observé par RPE les résidus insolubles au THF obtenus après différentes opérations d'hydroliquéfaction: avec ou sans catalyseurs, avec différentes concentrations d'un même catalyseur, avec un solvant donneur ou non donneur d'hydrogène, avec différentes pressions initiales d'hydrogène. Il nous a été possible de montrer (5) :
1) - que lors d'une opération d'hydroliquéfaction catalytique à 623K, l'hydrogène transféré au charbon provient essentiellement de la phase gazeuse: un phénomène d'"épandage d'hydrogène" sur le charbon a été avancé.
2) - que lors d'une opération d'hydroliquéfaction à 673K, l'hydrogène transféré au charbon et à ses produits de thermolyse provient, au moins pour une part, du solvant donneur d'hydrogène.
3) qu'il est possible de déterminer par des mesures RPE **ex situ** (sur les fractions insolubles au THF obtenues après hydroliquéfaction) un paramètre de RPE (dR°) qui reflète l'aptitude du milieu réactionnel à transférer de l'hydrogène au charbon et à ses produits de conversion. Ce paramètre permet également, dans certains cas, de détecter la présence de produits de repolymérisation dans les résidus d'hydroliquéfaction.

Le Tableau 2 résume les techniques utilisées et les informations qu'elles ont permis d'obtenir

Tableau 2

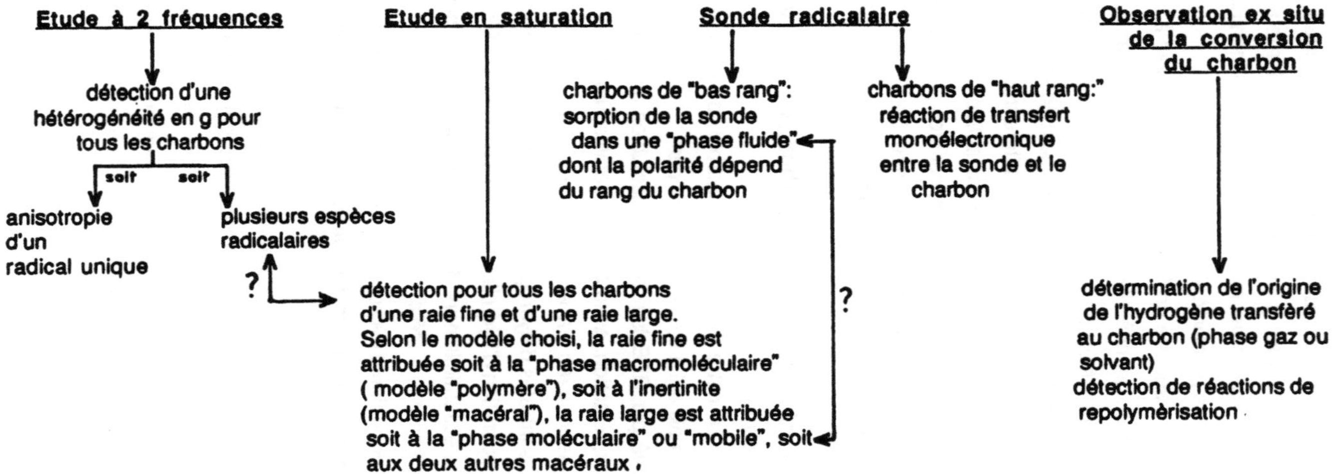

Etude à 2 fréquences
détection d'une hétérogénéité en g pour tous les charbons
soit
soit
anisotropie d'un radical unique
plusieurs espèces radicalaires
?
Etude en saturation
détection pour tous les charbons d'une raie fine et d'une raie large.
Selon le modèle choisi, la raie fine est attribuée soit à la "phase macromoléculaire" (modèle "polymère"), soit à l'inertinite (modèle "macéral"), la raie large est attribuée soit à la "phase moléculaire" ou "mobile", soit aux deux autres macéraux.
Sonde radicalaire
charbons de "bas rang": sorption de la sonde dans une "phase fluide" dont la polarité dépend du rang du charbon
charbons de "haut rang:" réaction de transfert monoélectronique entre la sonde et le charbon
?
Observation ex situ de la conversion du charbon
détermination de l'origine de l'hydrogène transféré au charbon (phase gaz ou solvant)
détection de réactions de repolymérisation

IV) Conclusion

L'étude par RPE de ce polymère particulier qu'est le charbon a donc permis d'obtenir des informations complémentaires de celles obtenues par d'autres méthodes en ce qui concerne sa structure, son degré d'évolution et sa réactivité à température ambiante vis-à-vis de certains réactifs (air, tanol). Elle a également permis d'évaluer les transferts d'hydrogène au cours de réactions d'hydroliquéfaction et de préciser si des réactions régressives de repolymèrisation avaient eu lieu ou non, ce qui est difficilement observable par d'autres méthodes.

Références

(1)J.UEBERSFELD,.A.ETIENNE,J.COMBRISSON
Nature, 1954, 174, 614.
(2) A.JEUNET, B.NICKEL, A.RASSAT
Fuel, 1989, 68, 883.
(3) A.JEUNET, B.NICKEL,A.RASSAT
Journal de Chimie Physique, 1988, 85, 97.
(4) B.NICKEL, A.JEUNET, A.RASSAT
in "Advanced Methodologies in Coal Characterization" (H.CHARCOSSET Ed. assisted by B.NICKEL-PEPIN-DONAT) Elsevier, Amsterdam 1990 p.149.
(5) B.NICKEL-PEPIN DONAT,A.JEUNET, H.CHARCOSSET, M.JAMOND
Fuel, 1990, 69, 856.

EPR Studies of Oxidation of Iron and Free Radicals Formation after Dessication of Oxyhemoglobin by Freeze-Drying and Spray-Drying

B. CHAILOT, B. CASSIER, P. LABRUDE*
Université de Bourgogne (Dijon) et
*Université de Nancy 1, France
Université de Bourgogne, Faculté de Pharmacie,
7 Bd J. d'Arc 21033 Dijon Cedex

Dans les Centres de Transfusion Sanguine, les hématies à péremption sont inutilisables. Elles peuvent être transformées en solution d'hémoglobine en vue d'une utilisation potentielle pour transfuser, mais cela suppose, pour ce type de produit sanguin que les problèmes de conservation résultant de l'oxydation de Fe(II) de l'hème soient résolus. Pour minimiser ce phénomène, on a voulu dessécher des préparations d'hémoglobines seules ou en présence de protecteur par différentes techniques à caractère industriel en vue d'étudier les produits formés au cours de ces traitements de dessiccation.

L'altération des globules rouges au cours de leur conservation se traduit par une oxydation de Fe(II) en Fe(III), et par là-même de l'oxyhémoglobine (HbO_2) en ferrihémoglobine (Hb^+). Nous avons cherché à caractériser les mélanges obtenus après la mise en oeuvre des traitements précités par différentes techniques, telles que la spectrophotométrie moléculaire visible ou U.V., le dichroïsme circulaire optique, l'électrophorèse en gel de polyacrylamide et la résonance paramagnétique électronique (R.P.E.) (1).

L'application de la R.P.E. à notre problème permet de caractériser la ferrihémoglobine (Hb^+) formée au cours des différents traitements mis en oeuvre. Hollocher (2) a étudié la dénaturation thermique de HbO_2 et mis en évidence une transition haut-spin ↔ bas-spin pour Hb^+ formée. Peisach (3) a montré que le changement de ligand sur l'ion Fe(III) des hémoprotéines s'accompagne d'une modification de la symétrie du champ cristallin dans lequel est plongé cet ion. Pour notre part, nous voulons préciser si la configuration de l'ion Fe(III) de Hb^+ formé par voie chimique diffère ou non de celle de Hb^+ formée après dessiccation par lyophilisation et nébulisation en faisant des mesures R.P.E sur la poudre, ou de Hb^+ formée après réhydratation de ces poudres en faisant des mesures R.P.E. sur la solution.

B. Catoire (Editor)
Electron Spin Resonance (ESR) Applications
in Organic and Bioorganic Materials

Les conditions de préparation des solutions d'oxyhémoglobine, les protocoles opératoires des traitements de lyophilisation, de nébulisation et de réhydratation ainsi que les conditions de mise en oeuvre des techniques autres que la R.P.E. sont décrits par ailleurs (1, 4, 5).

Les spectres R.P.E. sont réalisés au moyen d'un spectromètre VARIAN E 9, à la température de 77 K. Le tube porte-échantillon, en quartz et de diamètre intérieur 4 mm, obturé après remplissage, est plongé dans l'azote liquide pendant 30 minutes, puis placé dans le dewar à azote liquide et positionné dans l'entrefer de l'aimant. Les conditions de détection sont identiques pour tous les graphes montrés dans les résultats. Le paramètre déterminé expérimentalement est le facteur de décomposition spectrale g (6).

La figure 1 montre les tracés obtenus pour les solutions témoins d'oxyhémoglobine seule (courbe A) ou en présence d'un protecteur tel que le glucose (courbe B), le tris (courbe C) ou la β-alanine (courbe D). Ces spectres confirment que nos témoins ne renferment pas de Hb^{+}, et que les molécules protectrices ajoutées à des concentrations finales aussi élevées que 0,25 M ne provoquent pas de modification de la protéine.

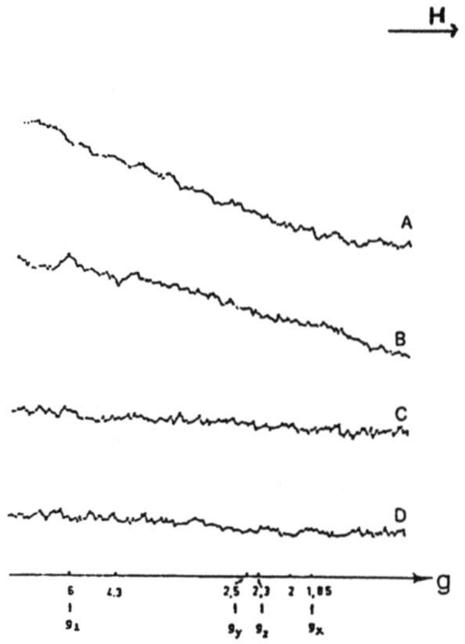

Figure 1: Spectres de Résonance Paramagnétique Electronique, avant lyophilisation, de HbO_2 seule (courbe A), de HbO_2 en présence d'α-D glucose (courbe B), de HbO_2 en présence de Tris (courbe C), de HbO_2 en présence de β-alanine (courbe D).

Les spectres de poudre obtenus après traitement de nébulisation sont semblables à ceux obtenus par lyophilisation (figure 2) et cela aussi bien pour l'oxyhémoglobine seule (courbe A) qu'en présence de glucose (courbe B), de tris (courbe C) ou de β-alanine (courbe D).

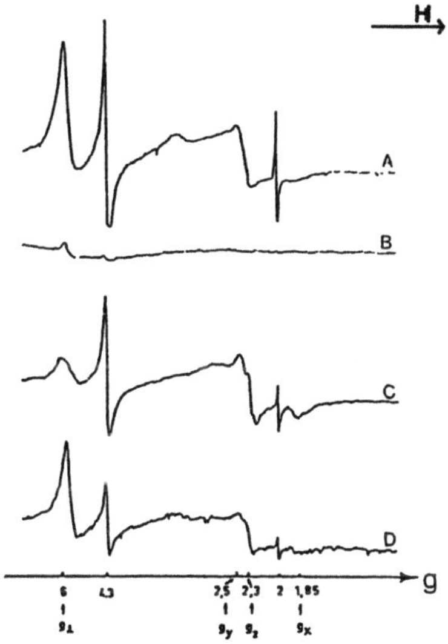

Figure 2: Spectres de Résonance Paramagnétique Electronique, après lyophilisation et avant réhydratation, de HbO_2 lyophilisée seule (courbe A), de HbO_2 lyophilisée en présence d'α-D glucose (courbe B) de HbO_2 lyophilisée en présence de Tris (courbe C), de HbO_2 en présence de β-alanine (courbe D).

L'interprétation de ces spectres nous permet d'affirmer que :

- le phénomène d'oxydation se produit au cours de la dessiccation ;

- la ferrihémoglobine formée présente la même configuration pour l'ion Fe(III). Or, partant dans les trois cas de solutions d'oxyhémoglobine à un pH donné, le fait d'obtenir le même spectre est un argument supplémentaire pour dire qu'il ne se produit comme dénaturation que l'oxydation de Fe(II) sans qu'apparaissent d'autres formes précipitées de l'hème telles que des hémichromes ;

- les protecteurs n'établissent pas de liaisons spécifiques au niveau de la poche de l'hème et leur mode d'action ne consiste pas en un blocage de l'ion Fe(II), car toutes les méthodes spectroscopiques montrent que les transitions spectrales de l'oxyhémoglobine seule (courbe A) ou en présence de protecteur sont identiques ;

- l'ensemble des raies des spectres A, C et D se décompose en quatre systèmes de résonance auxquels les valeurs de g sont attribuées par comparaison avec des standards ; c'est ainsi que la raie à g = 6 correspond à la composante g de l'absorption de Fe(III) haut-spin, sa largeur de raie à mi-hauteur étant de 100 gauss ; la raie à g = 4,3 est souvent présente dans les échantillons d'origine biologique et peut être absorbante et peut être due à des artéfacts de préparation car pouvant résulter d'espèces absorbantes situées dans un champ cristallin de symétrie mal définie ; la raie à g = 2,3 est la composante gz de l'absorption de Fe(III) bas-spin, donc de symétrie abaissée par rapport à la raie de g = 6 ; enfin la raie à g = 2,0, symétrique et de largeur à mi-hauteur inférieure à 20 gauss traduit la présence de radicaux libres, ces radicaux libres ayant été identifiés par De lamourd (7) par spin-trapping comme étant des anions surperoxyde O_2^-.

Les spectres obtenus après réhydratation sont représentés sur la figure 3.

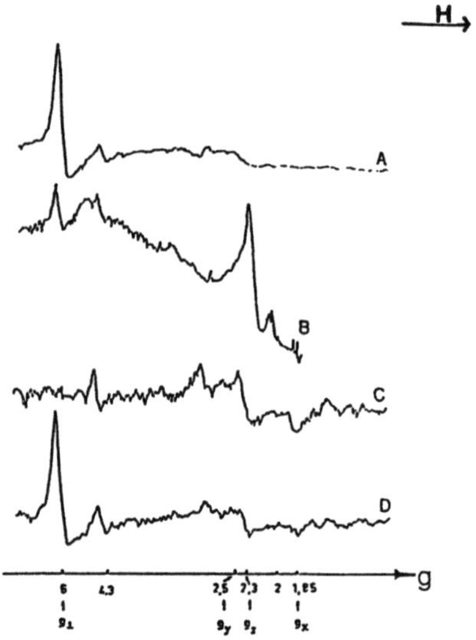

Figure 3: Spectres de Résonance Paramagnétique Electronique après lyophilisation et après réhydratation de HbO_2 lyophilisée seule (courbe A), de HbO_2 lyophilisée en présence d'α-D glucose (courbe B), de HbO_2 lyophilisée en présence de Tris (courbe C), de HbO_2 en prśence de β-alanine (courbe D.)

La détection R.P.E. en présence d'eau est perturbée par suite de l'absorption de l'énergie électromagnétique par le solvant et on ne peut pas comparer les amplitudes des spectres des poudres (figure 2) et des spectres des solutions (figure 3). Par contre, la comparaison entre deux des spectres de la figure 3 montre des variations du nombre et de la forme des raies pouvant correspondre à des dénaturations différentes induites au cours de la réhydratation. C'est ainsi que les spectres A et D possèdent une raie à $g = 6$ très intense alors que les spectres C et D ont des raies supplémentaires à $g = 4,3$ et 2,3. Ces résultats sont difficiles à interpréter, d'autant que la réhydratation des échantillons peut être différente et dépendre du protecteur présent à des concentrations aussi élevées.

L'ensemble de nos résultats suggère que les phénomènes d'oxydation produits ne diffèrent entre eux que d'un point de vue cinétique. Pour confirmer cette hypothèse, nous avons séché des solutions de HbO_2 par un courant d'air forcé, obtenant ainsi une vitesse de séchage plus faible qu'avec les autres traitements nous permettant de faire des mesures en cours de dessiccation. Après avoir vérifié par les techniques déjà citées que l'oxydation à l'air de HbO_2 ou auto-oxydation conduisait aux mêmes produits d'oxydation que les autres traitements, nous avons pu appliquer à notre problème les conclusions données dans la littérature concernant l'étude de l'auto-oxydation de HbO_2. En particulier, la présence de radicaux libres après lyophilisation, montrée par R.P.E., laisse à penser que les mécanismes d'oxydation de HbO_2 par séchage par voie chimique et par auto-oxydation sont identiques, et que l'oxydation est un processus radicalaire. Nous en avons déduit que pour éviter le phénomène d'oxydation, il fallait travailler en l'absence de dioxygène et proposer de lyophiliser de la carbomonoxyhémoglobine (HbCO) (8).

En conclusion, la R.P.E a permis de caractériser le produit d'oxydation de HbO_2 au cours de deux traitements à caractère technologique et de dire que les produits avaient une configuration identique. De plus, le configuration de Fe(III) dans ces composés est identique à celle du produit obtenu par voie chimique, laissant penser que le mécanisme d'oxydation est comparable. Ce résultat n'aurait pas pu être obtenu par une autre technique que la R.P.E., les mesures portant soit sur les solutions, soit sur les poudres. Nous avons pu montrer par ailleurs que le problème de l'oxydation de HbO_2 au cours de la dessiccation était d'ordre cinétique, la déshydratation accélérant l'oxydation de Fe(II) de la protéine.

REFERENCES BIBLIOGRAPHIQUES

(1) Chaillot B. (1984), Thèse Doctorat-ès Sciences, Université de Dijon

(2) Hollocher T.C., Buckley, L.M. (1966); J. Biol. Chem., 241, 1958-1968

(3) Peisach J., Blumberg, W.E., Ogawa, S. Rachmilewitz, E., A. Oltzik R., J. Biol. Chem., (1971), 246, 3342-3355

(4) Labrude P., Chaillot B., Vigneron C., (1987), J. Pharm. Pharmacol, 39, 344-348

(5) Labrude P., Rasolomanana, M. Vigneron, C. Thirion, C. Chaillot, B., (1989), J. Pharmi. Sci., 78, 223-229

(6) Dickinson L., C. Symons, M.C.R., (1983), Chem. Soc. Rew., 12, 387-414

(7) Delamourd L., Barrère B., Clerc Y., Saint-Blancard J., Letterier F. (1984), F. Trav. Scient. Serv. Santé Armées, 5, 193-196

(8) Chaillot B., Labrude P., Larcher D., Thirion C., Vigneron C., (1987), Colloque "Le globule rouge et autres transporteurs d'oxygène", Lyon, 17-19 Avril 1986, Collection Fondation Mérieux

CW and Pulsed EPR as a Tool for Studying the Structure of the Oxygen-evolving Enzyme of Photosystem II

Jean-Luc Zimmermann

Service de Biophysique, Dèpartement de Biologie, CEN de Saclay, 91191 Gif-sur-Yvette, France.

The Oxygen-evolving enzyme catalyzes the light-driven oxidation of H_2O to O_2 in Photosystem II of green plants chloroplasts. The enzyme is part of a membrane bound protein complex and contains four atoms of manganese and one or two Ca^{2+} ions. Considerable efforts are being made in many laboratories to unravel the structure of the metal cluster of this enzyme, which is of considerable biological importance. Five different oxidation states of the enzyme, S_0-S_4, can be described during its cycle, which apparently reflect the successive charge accumulation of the four positive charges that are necessary to oxidize 2 H_2O to one O_2 molecule. One of these states , S_2, is paramagnetic and gives rise to a multiline signal observable at liquid helium temperatures and bearing remarkable similarities with those of dimeric complexes of Mn or of higher nuclearities.

The EPR signal that arises from the S_2 state of the Oxygen evolving enzyme is shown in Figure 1A. This signal is completely broadened at room or liquid nitrogen temperatures and requires liquid helium temperatures (4-20K) to be observable. The signal (1) is centered at g=1.98 and presents a wealth of hyperfine structure (19-20 lines have been detected) over a spectral range of approximately 1800 Gauss. Such properties are typical of Mn complexes which have a S=1/2 ground state such as those having the central motif di-μ-oxo-Mn(III)Mn(IV) (Fig. 1C), with terminal ligands like e.g. bipyridine, phenantroline or triaminoethylamine (2,3). Recently a trimer of Mn: $[Mn(IV)_3O_4(bipyridine)_4Cl_2]$ has been synthesized and shown to display a multiline signal that also resembles that of the S_2 state of the Oxygen-evolving enzyme (4).

The structure of the biological complex is unknown, and in order to obtain some information from the spectroscopic properties of the S_2 signal, a

B. Catoire (Editor)
Electron Spin Resonance (ESR) Applications in Organic and Bioorganic Materials

theoretical study has been undertaken. Using a simple Heisenberg model, it was shown (5) that no complex of Mn having a nuclearity lower than four, and having reasonable oxidation states (i.e. III and IV (6)) can display an EPR spectrum that extends to 1800 Gauss, like the biological complex. The model used takes into account the fact that the EPR signal must arise from an S=1/2 total spin state, which is formed from combinations of the individual spin states of the ions that form the cluster (Mn(II), S=5/2, Mn(III), S=2, Mn(IV), S=3/2). A typical example of a simulation that results from this study is shown in Fig. 1B.

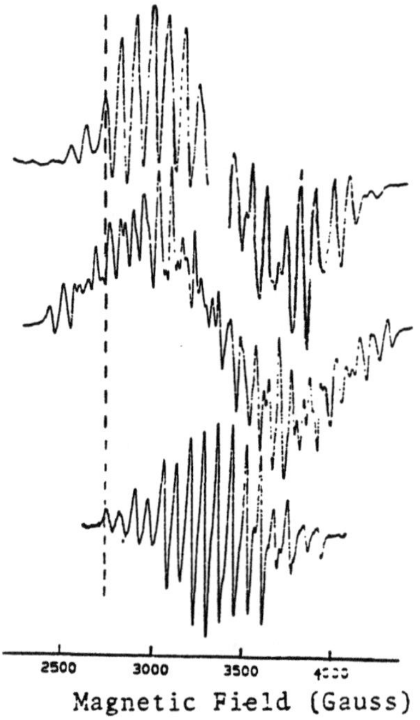

Fig. 1: A- The light induced EPR signal from the S_2 state of the Oxygen evolving enzyme measured at 8 Kelvin with a modulation amplitude of 20 Gauss. B- Simulation of the EPR spectrum of a Mn tetramer made of 3 MN(III) and one Mn(IV), resulting in a S=1/2 spin state. g=1.98 and the calculated hyperfine coupling constants are: A_1=132 G, A_2=81 G, A_3= 81 G, A_4=66 G. The calculated spectrum has a spectral width that is close to that originating from the biological complex. C- The EPR spectrum of Mn(III)Mn(IV)O_2(triaminoethylamine)$_2$ at 6 Kelvin and modulation amplitude of 5 Gauss (see (3)).

When Ca^{2+} is removed from the complex, the enzyme is no longer capable of oxidizing H_2O to O_2, yet it can still accumulate positive equivalents up to the S_3 state. It has recently been shown that in such a preparation, a new multiline signal is formed that corresponds to a modified S_2 state. This signal also arises from a ground S=1/2 state, but has a greater number of lines than the multiline signal observed in the functional enzyme (Fig.2) (7). From a study of the orientation dependence of these lines , it was concluded that anisotropy alone cannot explain the number of lines, and it was suggested that a cluster consisting of at least three Mn ions may be present as the origin of the multiline signal (8).

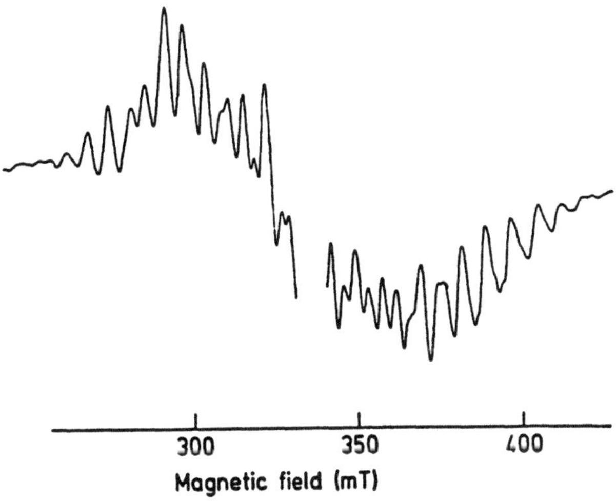

Fig. 2: The EPR signal from the S_2 state of the Oxygen evolving enzyme depleted of Ca^{2+} recorded under the same conditions as in Fig. 1A.

In addition, an EPR signal has been detected that arises from the S_3 state in the Ca^{2+} depleted enzyme. This signal is also observable at liquid helium temperatures, is centered at g=2.004 and is manifested as a rather broad (164 Gauss) and apparently derivative-like line. It was suggested that the S_3 signal could originate from an oxidized amino acid which interacts magnetically with the Mn cluster (6). Further charcterization of this state has produced the following results. Using an improved illumination protocol to produced the S_3 signal, it has been shown that appearance of the signal is accompanied by the disappearance of the S_2 multiline signal. The better resolved features of the signal at g=2.004 can be simulated assuming a weak magnetic exchange interaction (J=0.0063 cm^{-1}) between a free radical and a hypothetical S=1/2 Mn tetramer. In such a model, the observed signal arises from the superimposition of the four EPR allowed transitions in the quartet that results from the interaction between the two S=1/2 spins. The simulation also predicts the collapse of the hyperfine pattern of the multiline signal (Fig. 3). Judging from the near UV absorption spectrum of the S_2 to S_3 transition, which is remarkably similar to that of oxidized histidine, it was concluded that a histidine residue undergoes oxidation during this transition in the inhibited enzyme, and propose that this also occurs in the functional enzyme (7).

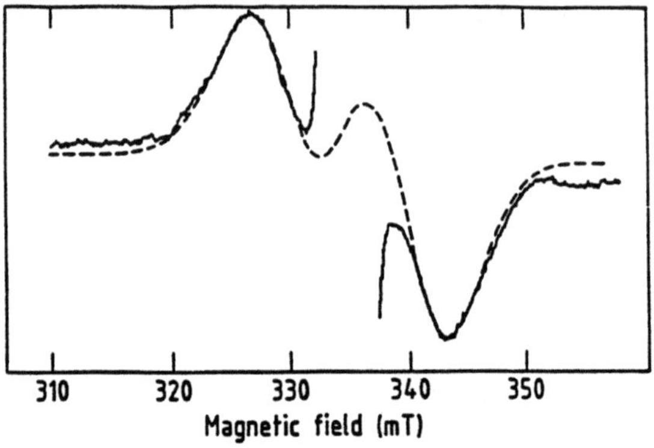

Fig. 3: The EPR signal of the S_3 state in the Ca^{2+} depleted enzyme (solid line), together with the calculation of the spectrum resulting from the interaction of a radical (S_1=1/2, g=2.012) and a Mn tetramer (S_2=1/2, g=1.965, A=152 G, A_2=76 G, A_3=53 G and A_4=49 G. A rather weak antiferromagnetic exchange coupling between the electronic spins (J=0.0063 cm^{-1}) was used to obtain the calculated spectrum.

NH_3 is known to inhibit the O_2 evolution reaction. One of the sites of action of this molecule has been suggested to be at the site of binding of the natural substrate water. When the O_2 evolving enzyme is treated with a low concentration (2mM free base) of NH_3, the S_2 state exhibits another form of multiline signal (Fig.4) (9). As with the other multiline signals observed in this system, attempts to simulate the exact shape of this signal have been unsuccessful; however, the question may be asked whether this modification of the multiline signal in the S_2 state reflects the binding of NH_3 to the Mn. A possible mechanism for this inhibition involves the binding of ammonia as Lewis base to the Lewis acid Mn, thereby competing for binding sites with a primary substrate, water (10). This question was addressed by means of a pulsed EPR method: Electron Spin Echo Envelope Modulation Spectroscopy (ESEEM).

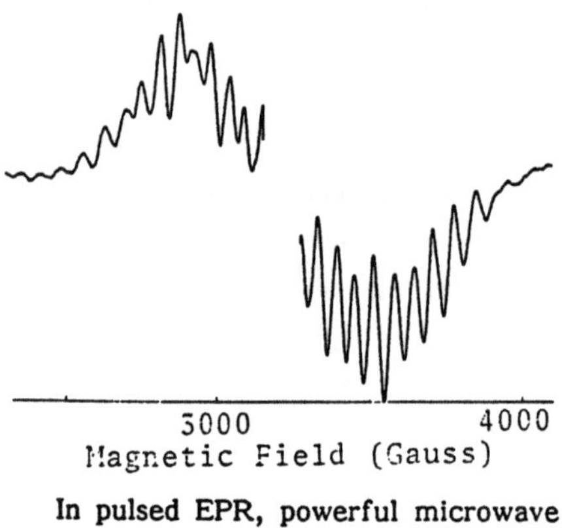

Fig. 4: The EPR signal of the 'relaxed' S_2 state of the enzyme inhibited with NH_3at pH=7.5. Instrument settings are as in Fig. 1A.

In pulsed EPR, powerful microwave pulses (typically a few hundred Watts) produce large rotations of the net electron spin magnetization on a time scale whic is short compared with the relaxation processes which return the magnetization to equilibrium. For EPR signals with large bandwidths, especially those associated with transition metals, only a small fraction of the inhomogeneously broadened line may be excited by a microwave pulse, and the FID may reflect only the excitation profile of the incident pulse and often decays to undetectable power levels before the instrument receiver recovers from the pulse. However, the rapid dephasing due to inhomogeneous broadening may be refocused with multiple pulse sequences, resulting in one or more observable electron spin echoes. In addition to inducing the electron spin transitions, the microwave pulses may also induce 'semiforbidden' transitions of paramagnetic nuclear moments magnetically coupled to the electron spins, resulting in quantum mechanical coherences in the nuclear spin sublevels associated with the electron spin levels. These coherences create interference effects which can be measured by varying the electron spin echo pulse timings. Fourier analysis of the resulting electron spin echo envelope modulation pattern reveals the frequencies of the various nuclear transitions. Typical examples of paramagnetic moments that may be coupled to electron spins in metal enzymes are protons (^1H, I=1/2), nitrogens from pyrroles and imidazoles (^{14}N, I=1, ^{15}N, I=1/2).

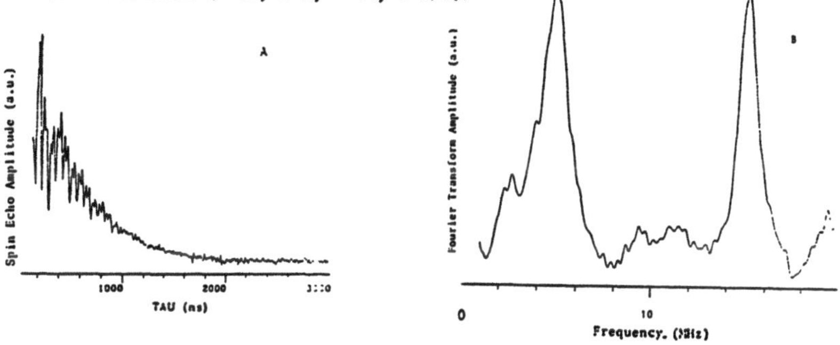

Fig. 5: Two pulse ESEEM (A) and Fourier cosine transform (B) of the S_2 state of the Oxygen evolving enzyme. The spectra were obtained at a temperature of 4.2 Kelvin, a microwave frequency of 9.2446 GHz, a magnetic field of 3500 G, and a time interval between successive spin echo pulse sets of 4.0 ms.

Fig. 5A shows the time domain ESEEM spectrum of a sample containing the O_2 evolving enzyme in the S_2 state. This spectrum is obtained by measuring the amplitude of the magnetization echo following two resonant microwave pulses, as a function of the time τ between the two pulses. The Fourier transform of this spectrum is shown in Fig. 5B: the peak at 14.9 MHz results from protons weakly coupled to the electron spin. A second large feature is observed at 4.7 MHz. The origin of this feature is still being characterized. The ESEEM spectrum of the enzyme treated with $^{15}NH_3$ is shown in Fig. 6. An additional low frequency feature can be observed at 3.12 MHz. It has been shown that this frequency arises from the coupling of the ^{15}N (I=1/2) magnetic moment. In such a system, the nuclear magnetic moment interacts with an effective magnetic field resulting from the summation of the applied magnetic field and the hyperfine field produced by the electron spin. In one of the electronic spin sublevels, the hyperfine and applied field add to produce a larger effective field (β manifold). In the α manifold, the hyperfine and applied field subtract to produce a smaller effective field. There are then two I=1/2 nuclear transitions that can be measured by ESEEM.

$$\nu_\alpha = \left| \nu_I - |A|/2 \right|, \qquad \nu_\beta = \left| \nu_I + |A|/2 \right| \qquad (1)$$

It was shown that the 3.12 MHz frequency corresponds to the ν_β of equation (1). At the observation field of 3500 Gauss, ν_I=1.51 MHz for ^{15}N, and the ^{15}N hyperfine coupling constant can be then calculated to be A(^{15}N)=3.22 MHz. The ν_α transition is not observed here because the frequency of this transition is very close to zero (ν_α=0.1 MHz).

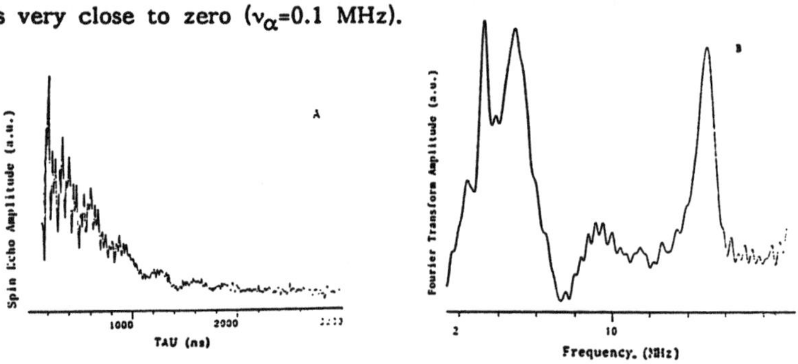

Fig. 6: Two pulse ESEEM (A) and Fourier cosine transform (B) of the 'relaxed' S_2 state of the enzyme treated with $^{15}NH_3$. The spectra were obtained as in Fig. 5.

The analysis of the ESEEM of the sample treated with $^{14}NH_3$ is complicated by the fact that ^{14}N (I=1) possesses an electric quadrupole moment which produces non degenerate nuclear sublevels even in the absence of an external magnetic field. However, the three pulse ESEEM pattern (not shown) could be fitted from resulting from an I=1 ^{14}N nucleus with A(^{14}N)=2.29 MHz, and with electric quadrupole parameters e^2qQ=1.61 MHz and η=0.59. The ^{14}N electric quadrupole parameters were interpreted with respect to possible chemical structures for the nitrogenous ligand. An amido (NH_2) bridge between two metal ions was proposed as the molecular identity of the ammonia-derived ligand to the catalytic Mn of the O_2 evolving enzyme (11).

Acknowledgements: I would like to thank my co-authors on papers which describe the work presented here: Jacques Bonvoisin, Alain Boussac, David Britt, Jean-Jacques Girerd, Mel Klein, Bill Rutherford and Ken Sauer. This work was supported in part by the CNRS (URA 1290).

References:

1. Dismukes, G.C. and Siderer, Y. (1980) FEBS Lett. 121, 78-80.
2. Cooper, S.R., Dismukes, G.C., Klein, M.P. and Calvin, M. (1978) J. Am. Chem. Soc. 100, 7248-7252.
3. Hagen, K.S., Armstrong, W.H. and Hope, H. (1988) Inorg. Chem. 27, 969-971.
4. Auger, N., Girerd, J.J., Corbella, M., Gleizes, A. and Zimmermann, J.L. (1990) J. Am. Chem. Soc. 112, 448-450.
5. Bonvoisin, J.J., Girerd, J.J. and Zimmermann, J.L., in preparation.
6. Guiles, R.D., Zimmermann, J.L., McDermott, A.E., Cole, J.L., Dexheimer, S.L., Britt, R.D., Wieghardt, K., Bossek, U., Sauer, K. and Klein, M.P. (1990) Biochemistry 29, 471-485.
7. Boussac, A., Zimmermann, J.L. and Rutherford, A.W. (1989) Biochemistry 28, 8984-8989.
8. Boussac, A., Zimmermann, J.L. and Rutherford, A.W. (1989) Proc. 8th Intern. Cong. Photosynthesis, Stockholm, in press.
9. Boussac, A., Zimmermann, J.L., Rutherford, A.W. and Lavergne, J. (1990) Nature, in press.
10. Velthuys, B.R. (1975) Biochim. Biophys. Acta, 396, 392-401.
11. Britt, R.D., Zimmermann, J.L., Sauer, K. and Klein, M.P. (1989) J. Am. Chem. Soc. 111, 3522-3532.

An ESR\ENDOR Study of Solvent Radical Adducts of 2,4,6-Tri-Tert-Butylnitrosobenzene (BNB)

C.L. Davies, J.C. Evans, C.C. Rowlands.

School of Chemistry and Applied Chemistry,

University of Wales College of Cardiff,

P.O. Box 912, Cardiff CF1 3TB.

M.J. Fildes, Shell Research LTD., Thornton Research

Centre, P.O.Box 1, Chester CH1 3SH.

Introduction.

Many common solvents are normally considered to be inert, i.e. playing no part in a reaction mechanism. However, this may not be strictly true when studying photolytic systems, an example being the kinetic study of hydrogen abstraction from hindered phenols by tert-butylperoxy radicals :-

$$tBuOO^{\cdot} + IH \xrightarrow{kI} tBuOOH + I^{\cdot}$$

where tBuOO· is formed from photolysing a mixture of di-tert butyl peroxide and tert-butylhydroperoxide :-

$$tBuO\text{-}OtBu \xrightarrow{h\nu} 2\ tBuO^{\cdot}$$

$$tBuO^{\cdot} + tBuOOH \longrightarrow tBuOH + tBuOO^{\cdot}$$

In such a case, it is possible for side reactions that involve the generation of solvent radicals and their subsequent reactions to become an important factor, effectively masking the measurement of the true kinetics.

B. Catoire (Editor)
Electron Spin Resonance (ESR) Applications in Organic and Bioorganic Materials

The object of this work has been to investigate the behaviour of certain organic solvents with respect to solvent radical generation, in the presence of hydrogen abstractor precursors, such as di-tert-butyl peroxide and benzophenone (1,2,3). The spin trapping technique (4) was used to study the short-lived radical species, with BNB as the trap.

Terabe and Konaka (5) initially recognised the advantages of using BNB as a spin trap. They identified the major benefits as being (a) its stability toward UV irradiation (hence making it a useful probe in the study of photochemical systems), and (b) the presence of two radical trapping sites (the nitrogen and oxygen atoms of the nitroso group).

The latter advantage provides an insight into the actual steric properties of the trapped radical species, as bulky tertiary alkyl radicals usually prefer type 2, i.e. oxygen addition (6), whereas primary alkyl radicals tend to prefer type 1, i.e. nitrogen addition (6). Secondary alkyl radicals however can undergo either addition, possibly to give a mixed adduct spectrum (6), in which the adduct ratios will depend on the overall structure of the radical.

The two types of adducts are readily distinguishable by their g values and nitrogen splitting constants, with typical values being :

	Type 1	Type 2
aN/Gauss	12 to 14	approx. 10
g	2.006	2.004

Experimental.

Materials.

The following chemicals were supplied by Aldrich Chemical Co. and used without further purification: cyclopentane, cyclohexane, methylcyclopentane, methylcyclohexane (HPLC grade solvents), tetrahydrofuran-d^8, cyclohexane-d^{12}, cyclopentyl bromide, cyclohexyl bromide, iodomethane (stabilised with copper), hexamethylditin, 2,4,6-tri-tert-butylnitrosobenzene. The following chemicals were also used as received : di-tert-butyl peroxide (<5% tert-butyl hydroperoxide) supplied by Fluka; benzophenone supplied by BDH. Tetrahydrofuran was distilled under N_2 with cuprous chloride, sodium hydroxide pellets and dried over sodium wire.

Procedure.

From stock solutions of ROOR (approx. 0.3 ml), BNB (approx. 6mg) and solvent (approx. 6ml), anaerobic samples were prepared and directly photolysed at room temperature, in the cavity of a Varian E3 spectrometer (either a Bausch and Lomb 150W Xe lamp or a medium pressure Hg lamp may be used). ENDOR experiments were carried out in the temperature range 193 to 223 K using a Varian E109 spectrometer, interfaced with a Bruker digital ENDOR resonance attachment and with a modified Oxford Instruments variable temperature controller. Similar experiments with anaerobic solutions of solvent, BP and BNB were photolysed as previously described.

Hexamethylditin (under nitrogen) was added to nitrogen-purged solutions of BNB and cyclohexyl bromide, cyclopentyl

Table 1. Hyperfine Coupling Constants for the BNB adducts of THF and THF-d^8.

(a) Using ROOR.

Solvent	ESR/Gauss			ENDOR/Gauss	g
	aN	a(m-2H)	a(α-H)		
THF	11.24	1.85	0.99	1.9(m-2H) 1.04(α-H) 0.162(tBu)	2.0041
	12.1				2.0075
THF-d^8	11.3	1.85			2.0043
	11.98		0.74		2.0068

(b) Using BP.

Solvent	ESR/Gauss			g
	aN	a(m-2H)	a(α-H)	
THF	11.24	1.85	0.98	2.0042
	12.1			2.0075
THF-d^8	11.3	1.85		2.0041
	11.98		0.74	2.0065

bromide and the solutions evacuated and photolysed as previously described. For control purposes, anaerobic solutions of BNB with ROOR, and BNB with iodomethane were photolysed as previously described.

For all cases, g values for the adducts were measured with respect to an α,α'-diphenyl-β-picrylhydrazyl (DPPH) marker.

Results.

Figures 1a,1c,1d and 1f show the ESR and ENDOR spectra of the radical adducts of BNB, ROOR, tetrahydrofuran and tetrahydrofuran-d^8, respectively. It is clearly seen from Figure 1f that a pair of proton peaks are missing from the ENDOR of the deuterated solvent adduct spectrum (cf. Figure 1c). From Figure 1a, the resonance marked (*) is seen to be resolved in Figure 1d. Hyperfine coupling constants are given in Table 1.

Figures 2a and 2c show examples of ESR and ENDOR spectra of radical adducts of BNB, ROOR and cyclopentane. Similar spectra were observed for cyclohexane. In the case of cyclohexane-d^{12} ENDOR, one pair of proton lines (at 11.4 and 16.5 MHz) were observed. Hyperfine coupling constants are given in Table 2.

Figures 3a and 3c show ESR and ENDOR spectra of the radical adducts of BNB, ROOR and methylcyclopentane. The ENDOR spectrum of Figure 3c shows the presence of two radical species with varying concentrations. Hyperfine coupling constants are given in Table 3.

Figures 1b, 1e, 2b, 3b, show computer simulations of the ESR spectra, based on values given in the respective tables.

All radical adducts observed were stable in the absence of UV, with no radical species being detected prior to irradiation.

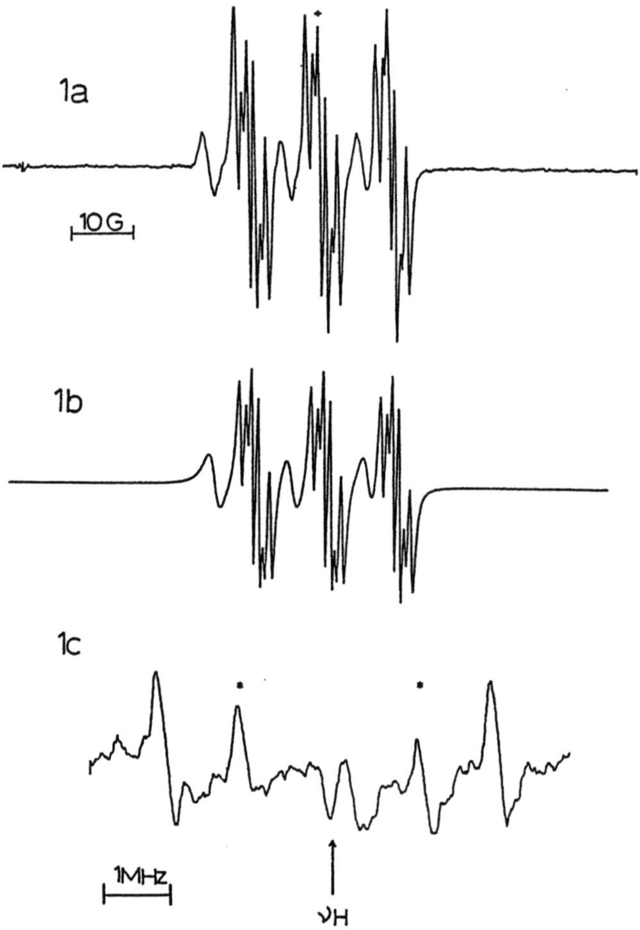

Figure 1a. The ESR spectrum of the THF adduct of BNB, formed during anaerobic photolysis with ROOR at room temperature.

Figure 1b. Computer simulation of the ESR spectrum in Figure 1a, using values given in Table 1.

Figure 1c. ENDOR spectrum of the THF adduct at 196 K . A total of 192 scans, accumulated by computer.

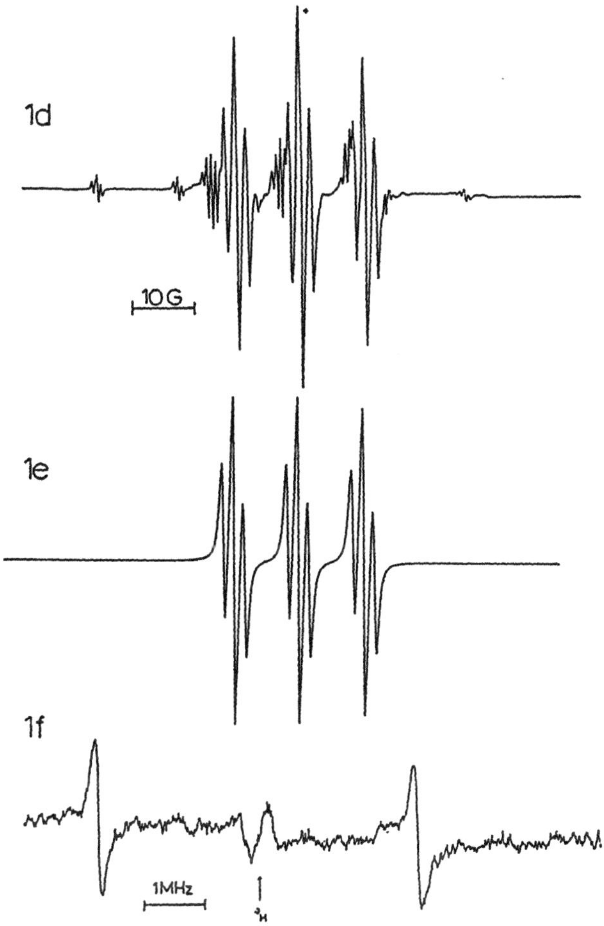

Figure 1d. The ESR spectrum of the THF-^{8} adduct of BNB, formed during anaerobic photolysis with ROOR at room temperature.

Figure 1e. Computer simulation of the THF-d^{8} adduct, using values given in Table 1, with one sixth of the proton coupling used for the deuterium coupling.

Figure 1f. ENDOR spectrum of the THF-d^{8} adduct, at 198 K.

Table 2. Hyperfine Coupling Constants for BNB solvent adducts.

Using ROOR.

Solvent	ESR/Gauss aN	ENDOR/Gauss	g
Cyclopentane	10.8	2.083(m-2H) 1.845(α-H) 0.179(tBu of trap)	2.0043
Cyclohexane	10.95	1.87(m-2H) 1.59(α-H) 0.22(tBu of trap)	2.0042
Cyclohexane ($d-^{12}$)	10.9	1.88(m-2H)	2.0043

Discussion.

Tetrahydrofuran.

From the hyperfine coupling constants in Table 1, it is seen that irrespective of the mode of solvent hydrogen abstraction, the same radical species is trapped.

Analysis of the ESR spectrum (Figure 1a) reveals two radical adduct species with quite different g values, and constitutes a mixed adduct spectrum. Proton ENDOR observed from sitting on the ESR line marked (+) in Figure 1a revealed three proton couplings, as shown in Figure 1c. The smallest coupling is assigned to the tert-butyl groups of BNB and an unambiguous assignment of the two remaining proton resonances was made by using tetrahydrofuran-d^8. Proton ENDOR experiments were carried out on the stable deuterated system, sitting on the ESR line marked (+) in Figure 1d. This time (Figure 1f), only two proton resonances were observed, which could be assigned to the tert-butyl groups. The two meta protons of the trap (with the absorptions marked (*) in Figure 1c) are now seen to be absent and confirm the presence of a_deuterated solvent adduct. No deuterium ENDOR was observed.

The remaining proton resonance in the tetrahydrofuran case was assigned to the alpha proton of a spin- trapped tetrahydrofuranyl radical.

For THF, the adduct with g $\simeq$ 2.0042 is identified as the N-alkoxyanilino radical formed by type 2 addition of the tetrahydrofuranyl radical to BNB (Adduct 1), formed by abstraction of an alpha proton from a THF solvent molecule by tert-butoxy radicals, or the triplet state 1,2-diradical of benzophenone (8).

Adduct 1.

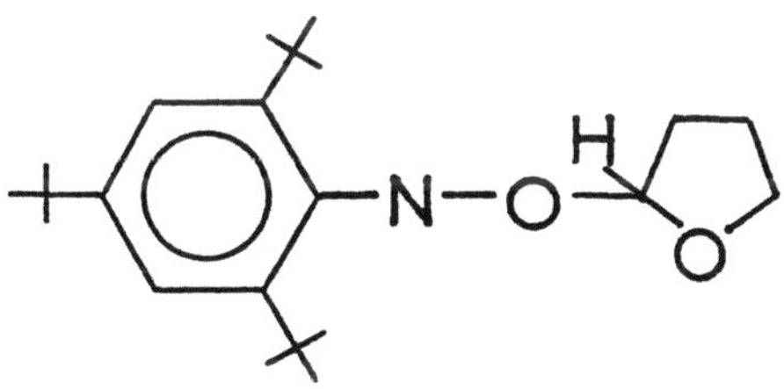

It is observed that the triplet arising from the two meta protons of the trap is further split into doublets by the alpha proton of the trapped species.

For THF-d^8, the ESR spectrum again indicates the trapping of two radical species with different g values (Figure 1d). The adduct with g $\simeq$ 2.0042 is similarly identified as the trapping of

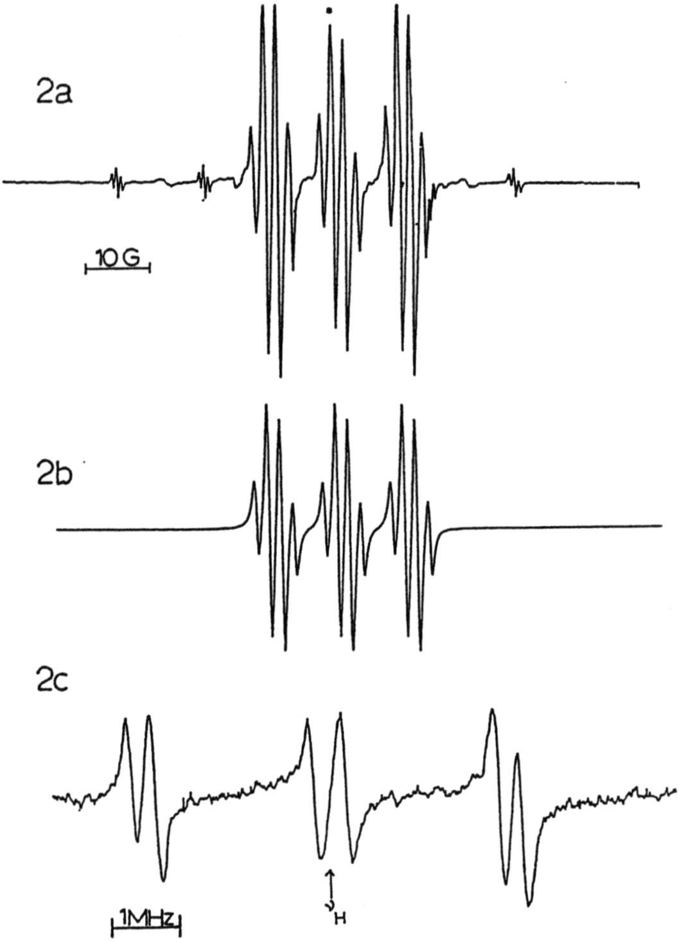

Figure 2a. The ESR spectrum of the cyclopentyl adduct of BNB, formed during anaerobic photolysis with ROOR at room temperature.

Figure 2b. Computer simulation of the ESR spectrum of the cyclopentyl radical adduct shown in Figure 2a, using values given in Table 2.

Figure 2c. ENDOR spectrum of the cyclopentyl adduct at 202 K.

the deutero-tetrahydrofuranyl radical, formed as before. In this case, only the triplet from the trap's two meta protons is observed. As aD $\simeq$ aH/6.5, super-hyperfine from the alpha deuterium of the THF radical is not observed as it falls within the ESR linewidth (ΔH = 0.5G). This was confirmed by computer simulation.

The mechanism by which hyperfine couplings from the proton on the tetrahydrofuran moiety is observed, can be explained by both hyperconjugation and spin polarisation, with the latter playing a more important role in this situation (see later discussion).

The radical adduct with g $\simeq$ 2.007 is common to both spectra (Figures 1a and 1d), suggesting type 1 addition. In the THF case, the spectrum consists of an unresolved triplet, whereas in the THF-d^8 case, line sharpening and resolution is observed. This species, however, remains uncharacterised.

Cyclopentane and Cyclohexane.

From the hyperfine coupling constants given in Table 2, it can be seen for each independent solvent (irrespective of the mode of hydrogen abstraction) that the same types of radicals are trapped.

The ESR and ENDOR spectra obtained for both cyclopentane (Figures 2a, 2c) and cyclohexane showed similarities. For both solvents, the ESR spectra are analysed as containing one major radical species, with the hyperfine splitting parameters and g values indicating type 2 addition. For cyclopentane, proton ENDOR observed from sitting on the ESR line marked (*) in Figure 2a, revealed three proton couplings (Figure 2c), with the smallest assigned to the tert-butyl groups of the spin trap.

Table 3. Hyperfine Coupling Constants for BNB Solvent Adducts.

(a) Using ROOR.

Solvent	ESR/Gauss aN	ENDOR/Gauss	g
Methyl-	10.5	1.87(m-2H)	2.0055
cyclopentane	10.0	2.14(m-2H)	2.0043
		0.2(tBu of trap)	
Methyl-	10.12		
cyclohexane	10.25	2.00(m-2H)	2.0040
		0.2(tBu of trap)	

(b) Using BP.

Solvent	ESR/Gauss aN	ESR/Gauss a(m-2H)	g
Methylcyclopentane	10.5	1.85	
	10.0	1.85	2.0042
Methylcyclohexane	10.1	1.82	2.0051
	10.35	1.85	2.004

Ambiguity of the assignment of the remaining two proton resonances was clarified using cyclohexane-d^{12}. From the hyperfine parameters given in Table 2 for the deuterated solvent, we can again infer that the same radical species was trapped. Proton ENDOR gave only one proton resonance, which was assigned to the meta protons of the trap.

Depending on solvent, the adduct is identified as that arising from the trapping of the cyclopentyl, cyclohexyl, or cyclohexyl-d^{11} radicals (Adduct 2).

Adduct 2.

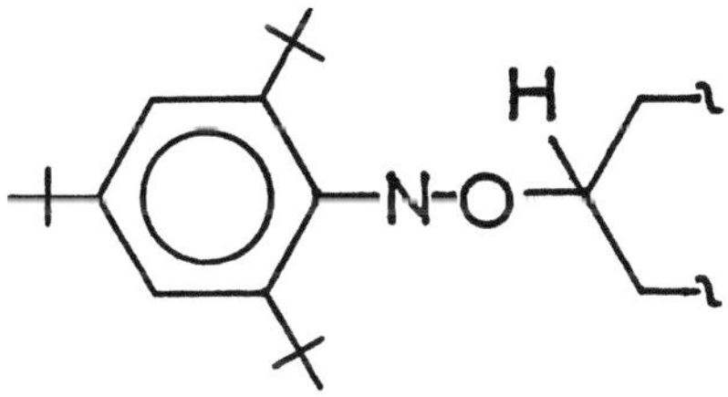

Terabe and Konaka explained the observation of a quartet in the ESR spectrum for the radical adduct of cyclohexane, as arising from the hyperfine coupling of the alpha proton of the trapped species being equal in magnitude to that of the meta hydrogens of the trap (6). Similarly, this argument may be extended to the radical adduct spectrum for cyclopentane (Figure 2a). However, the use of ENDOR has served to clarify that the alpha proton coupling in cyclopentane is not exactly the same as the trap's meta proton coupling. The same observation was made for cyclohexane, thus showing Terabe and Konaka's interpretation to be incorrect. However, one should note that the ESR adduct spectra can be computer simulated, either on the basis of Terabe and Konaka's interpretation or using the more correct ENDOR values. This shows that the simulation of spectra need not neccesarily lead to an unambiguous assignment of radical structures.

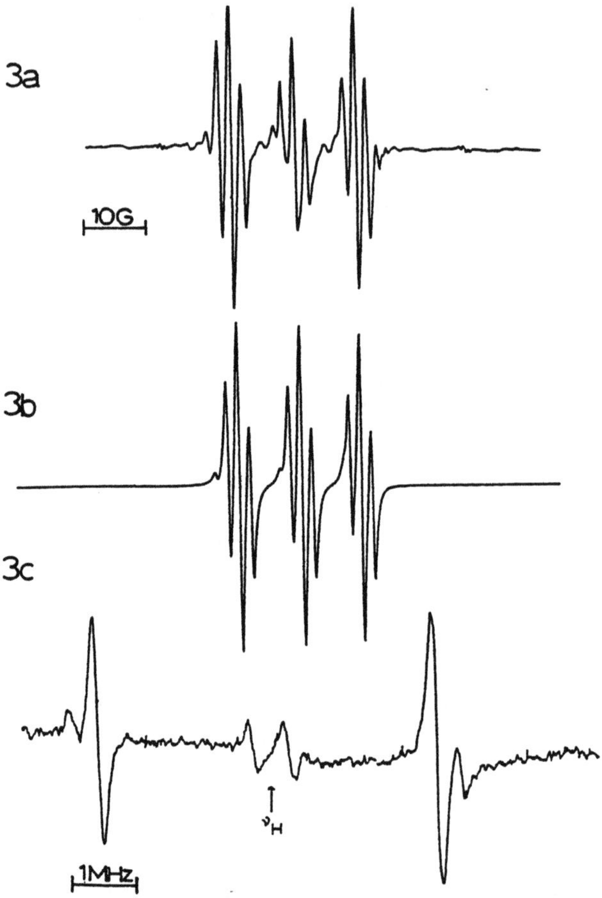

Figure 3a. The ESR spectrum observed for methylcyclopentane, on anaerobically photolysing with BNB and ROOR at room temperature.

Figure 3b. Computer simulation of the ESR spectrum of Figure 3a, using values given in Table 3, based on ratio of 1:10 for the radical adducts.

Figure 3c. ENDOR spectrum arising from the trapping of two radical species with varying concentrations, on the anaerobic photolysis of methylcyclopentane with BNB and ROOR at 190 K.

For the deutero case, the deuterium coupling is approximately aH/6.5, hence deuterium hyperfine is contained within the ESR linewidth, giving rise to a triplet spectrum. This verifies the observation of a deuterated cyclohexyl adduct and hence a cyclohexyl adduct.

In contrast to the THF case, the hyperconjugation mechanism must now be predominant, owing to the absence of the oxygen atom present in the THF case, thus reducing the spin polarisation effect. All of the radical adducts are confirmed by computer simulation.

Further experiments were carried out using hexamethylditin and the corresponding 1-bromo compounds of cyclopentane and cyclohexane. Photolysis of organotin compounds like hexamethylditin ($(CH_3)_3Sn_2(CH_3)_3$), results in the generation of trimethyltin radicals, which reduce alkyl halides to the corresponding alkyl radicals (7):-

$$R_3Sn\text{-}SnR_3 \xrightarrow{h\nu} 2\ R_3Sn^{\cdot}$$

$$R_3Sn^{\cdot} + R'\text{-}X \longrightarrow R_3Sn\text{-}X + R'^{\cdot}$$

Photolysis of the bromo solvents with BNB under vacuum gave no radical species, whilst in the presence of the tin compound under similar conditions, ESR spectra were obtained typical of the trapping of cyclopentyl and cyclohexyl radicals.

Methylcyclopentane and Methylcyclohexane.

From the hyperfine coupling constants for each solvent (Table 3), it can be seen that the same types of radicals were trapped.

Figures 3a and 3c show the ESR and ENDOR spectra observed for methylcyclopentane. For each solvent, the ESR spectra can be analysed as containing two main radical species with similar nitrogen and meta proton couplings, but different g values, the magnitudes of which indicate type 2 addition. From both Figures 3a and 3c, it is seen that the radical species are present in differing concentrations. In proton ENDOR experiments, three proton couplings were observed, with again the smallest assigned to the tert-butyl groups of the trap. For methylcyclopentane, ENDOR illustrates the presence of two radical species in varying proportions. By comparison, with methylcyclohexane, ENDOR illustrates the presence of a more dominant radical, with the lesser dominant species, we believe, falling within the linewidth of the ENDOR spectrum.

The ESR spectra are analysed as consisting of two overlapping triplets, with no hyperfine being observed from the trapped species. This is in contrast to the case for cyclopentane, where a hyperfine coupling of 1.845 gauss was observed. We attribute this to steric effects and where neither hyperconjugation nor spin polarisation mechanisms operate.

The two adducts can be accounted for by the trapping of the following radicals:- $(\dot{C}_6H_{10})CH_3$,$(C_6H_{11})\dot{C}H_2$; $(\dot{C}_5H_8)CH_3$,$(C_5H_9)\dot{C}H_2$. The first radicals of the above pairs are expected to dominate as their formation is via abstraction of a more labile proton, than that of a methyl proton, to form the tertiary radical. The primary radicals would be expected to undergo type 1 addition, but their non-planar conformations afford bulky radicals and, together with the bulky tert-butyl groups of the trap, we believe that these steric grounds would force them to undergo type 2 addition. Interestingly, we have observed the BNB nitroxide adduct (also found by Terabe and Konaka (6)) arising from the trapping of toluene radicals ($PhCH_2$). The planarity of the aromatic ring affords a less bulky radical, which permits type 1 addition. The presence of two radical species was verified by computer simulation.

Common to all three categories of solvents investigated is the control reaction of BNB and ROOR. Under anaerobic conditions, photolysis gave a completely different adduct, which was assigned to the trapping of a methyl radical and arose from the beta scission of tert-butoxy radicals under photolytic conditions (1,2). We believe the minor species present in Figure 2a is also this species. Verification of this adduct was made by photolysing iodomethane with BNB anaerobically, where methyl radicals were again trapped.

Conclusions.

For a variety of organic solvents, solvent radicals, which result from hydrogen abstraction by tert-butoxy radicals and the benzophenone triplet, have been successfully trapped by BNB and characterised using ESR and ENDOR techniques.

Furthermore, both the benzophenone triplet and tert-butoxy radicals react in the same manner.

In most cases, radical verifications have been carried out using deutero solvents or by reducing the corresponding organo-bromo compounds with hexamethylditin.

Acknowledgements.

This work is dedicated to the memory of the late Dr. Jeff Evans. C.L.D. is grateful to Shell Research Ltd., for funding her Ph.D studentship.

References.

1. Free Radicals. W.A. Pryor. Mc Graw-Hill Inc., 1966.
2. Photodegradation, Photo-oxidation and Photostabilisation of Polymers. B. Ranby, J.F. Rabek. John Wiley and Sons Ltd., 1975.
3. Mechanisms of Photophysical Processes and Photochemical Reactions in Polymers, Theory and Applications. J.F. Rabek, John Wiley and Sons Ltd., 1987.
4. E.G. Janzen, Accounts Chem. Res., 1971, 4, 31-40.
5. S. Terabe, R. Konaka, J.A.C.S., 1971, 93, 4306-4307.
6. S. Terabe, R. Konaka, J.C.S. Perkin Trans. (II), 1973, 369-374.
7. A.G. Davies, Chapter 9; Chemistry of Tin, ed. P.G. Harrison, Blackie and Sons Ltd., 1989.
8. E.D. Owen, I. Pasha, J. App. Polym. Sci., 1980, 25, 2417-24.

Comparative ESR Spin Trapping Study of Gamma Induced Radicals in Some Mono- and Oligosaccharides: Multicomponent ESR Spectra Simulation with the VOYONS Program

C.L.THIERY, J.B.TRIOLET, J-P.L.AGNEL, C.M.BATTESTI, P.B.VINCENT, J.J.RAFFI*
Laboratoire de Radiochimie des Constituants des Aliments
and J.M.THIERY
Modélisation et Génie Logiciel

Département de Physiologie Végétale et Ecosystèmes,
Centre d'Etudes Nucléaires de Cadarache, 13108 StPaul-lez-Durance, FRANCE.

INTRODUCTION

The radiolysis mechanism of oligosaccharides has been extensively studied during the last years with the Electron Spin Resonance (ESR) technique [1-4]. We report here experiments carried out using the spin trapping method [5,6] on α-D-glucopyranose, ß-D-glucopyranose, 1-O-CH_3-α-D-glucopyranose, sucrose and oligosaccharides Gn (defined as O-α-D-glucopyranosyl-$(1\text{->}4)_{n-1}$-α-D-glucopyranose) with n=2 to n=7, and several ^{13}C labeled derivatives of these compounds. The gamma induced radicals $S^\bullet$ are converted into long-lived nitroxide spin adducts in the liquid phase (water-ethanol [2:1]) by reaction with 2-methyl-2-nitrosopropane (MNP or tBu-N=O):

$$S^\bullet \; + \; tBu\text{-}N{=}O \; \text{--->} \; tBu\text{-}\underset{\displaystyle O^\bullet}{\underset{|}{N}}\text{-}S$$

In typical experiments, numerous radicals are trapped: their ESR signals show only slightly different hyperfine splitting constants (a_N and a_H) and very similar g-factors: consequently, they are all superimposed. The analysis of such ESR spectra is very complex. A partial separation of the mixture of nitroxide sugar radicals is obtained with High Performance Liquid Chromatography (HPLC) [7,8]. A micro-computer analysis of the ESR spectra is performed, with a specially adapted version of the program *VOYONS* [9,10] dedicated to the simulation of multicomponent ESR spectra in solution.

Correlations between radicals radioinduced in one sugar and its specifically labeled derivatives are attempted. They provide new informations on the chemical structure of the radicals induced in gamma irradiated α-D-glucopyranose, its oligomers G2 to G7 and sucrose.

B. Catoire (Editor)
Electron Spin Resonance (ESR) Applications in Organic and Bioorganic Materials

MATERIALS AND METHODS.

Sugars were purchased from Aldrich, Boehringer Mannheim, Fluka, Merck and Prolabo Companies, and MNP from Aldrich. The labeled compounds were synthesized in the Laboratory.

The sugars were irradiated in a ^{137}Cs cell, which supplies a dose rate of 3.4 $kGy.h^{-1}$. The applied doses (20 kGy for irradiation in the solid state at 77°K or ~298°K, and 2 kGy for irradiation in the liquid phase at ~273°K) provided intense enough ESR spectra. One minute before trapping, one volume of a 12 mg per mL solution of MNP in deoxygenated ethanol was added to two volumes of deoxygenated water.

- In the case of irradiation in the solid phase, the sugar was then dissolved in this water-ethanol [2:1] solution of MNP (~60mg/mL).
- In the case of irradiation in the liquid phase, the sugar was dissolved in an aqueous solution of MNP (stirred overnight), irradiated and, after irradiation, in order to prevent the formation of radicals derived from the alcohol, added to ethanol.

In some experiments, the ESR spectra were directly recorded. Most often, the spectra were too complex to be analyzed, owing to the superposition of numerous signals. Therefore, a partial separation of the spin trapped radicals was achieved with the use of a HPLC column. In that case, the nitroxide solution (2 mL) was immediately injected onto a Waters semi preparative C18 micro-bondapak column, cooled to ~278°K in order to increase the radicals stability. The water-ethanol [2:1] eluent was delivered at a flow rate of 1 $mL.min^{-1}$ and the ESR spectrometer was used first as a detector. The collected fractions (one per minute) were stored in an ice-bath until spectra recording.

The ESR spectra were recorded with an ASPECT 2000 computer, linked to a Bruker ER 200 D 10 ESR spectrometer. Simulations were performed on a Deskpro Compaq 386/25 micro-computer with the *VOYONS* program [9,10]. *VOYONS* is a general interactive simulation program, written in Turbo Pascal for IBM compatible 16-32 bit micro-computers.

A typical analysis of an ESR spectrum with the *VOYONS* program.

The informative part (from channel 340 to 1708) of the 2k spectrum recorded on the ASPECT 2000 is transferred, after baseline correction, to the micro-computer. The spectrum is systematically centred at channel 1024, corresponding to g = 2.0060, on the basis of the experimental measurements of the frequency, central field and sweep width.

In most cases, the spectrum shows several superimposed components. The probable number of species and approximate values for their ESR parameters, i.e. hyperfine splitting constants a_N, a_H, a_{13C}, linewidth and position of the spectral centre (g factor), are deduced from a careful observation of a series of related spectra, such as the different fractions of an HPLC experiment.

These values are given as starting guesses in a modified SIMPLEX [9-11] procedure. The fitting can be restricted to a specific zone (i.e. between channels 400 and 1200). For each set of parameters, the deviation spectrum (experimental spectrum minus simulation one) shows the quality of the fit; an objective evaluation is given by the computed standard deviation (error) of the deviation spectrum.

The program performs a systematic investigation until the best simulation (with the lowest error) is found. When the spectra of the different fractions of an HPLC experiment are simulated, the coherence between the computed parameters must be checked; the total (i.e. recorded before HPLC) ESR spectrum fitting can then be achieved.

Presently, the computer fit allows the detection of up to 6 radicals. As an exemple, Figure 1 shows the experimental, simulation and deviation spectra of G3, irradiated in the liquid phase; for each component, the hyperfine splitting constants and the linewidth are printed in mT; the g-factor and the coefficient (contribution) are indicated.

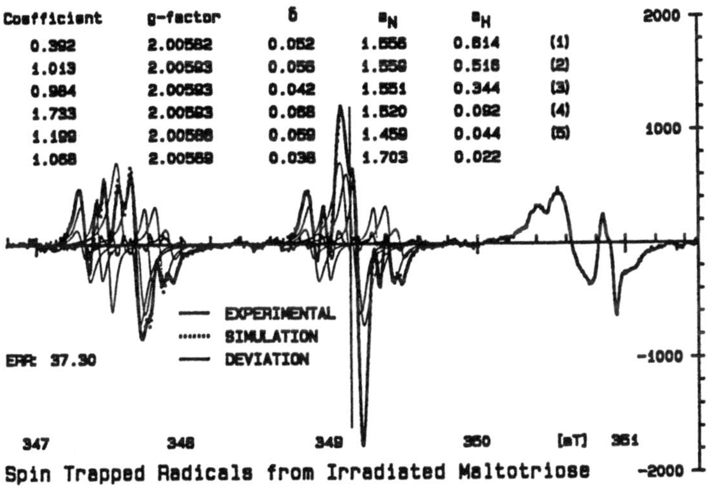

Figure 1

ESR spectrum of the nitroxide sugar radicals from O-α-D-glucopyranosyl-(1->4)-O-α-D-glucopyranosyl-(1->4)-α-D-glucopyranose, (G3), gamma irradiated (~2 kGy) at ~273°K, in an aqueous solution of 2-methyl-2-nitrosopropane (MNP); ESR recording is performed at room temperature in water-ethanol [2:1] solvent. Experimental, simulation and deviation spectra are displayed; ERR: value of the standard deviation. The vertical line indicates the g = 2.0060 position. On top, the individual parameters (coefficient, g-factor, linewidth and splitting constants a_N and a_H, expressed in mT) are printed for the nitroxide sugar radicals (1), (2), (3), (4) and (5); the 6th component is due to the spin trap.

TABLE 1

ESR hsc			Sugars					
a_N (mT)	a_H (mT)	a_{13C} (mT)	αG	ßG	[^{13}C-1]αG	[^{13}C-2]αG	[^{13}C-4]αG	[^{13}C-6]αG
(1)								
1.5	.11	-	+	+	+	+	+	-
(1)*								
1.5	.10	.46	-	-	-	-	-	+
(2)								
1.54	.45	-	+	+	+	+	+	-
(2)*								
1.54	.45	.42	-	-	-	-	-	+

ESR hsc: ESR hyperfine splitting constants: a_N, a_H, a_{13C} are expressed in mT (1mT = 10G); accuracy: ± 0.01;
αG: α-D-glucopyranose;
ßG: ß-D-glucopyranose;
In the following columns, + signs (resp. - signs) indicate the presence (resp. the absence) of the corresponding radical.

RESULTS AND DISCUSSION.

Assignment of two main common radicals in monosaccharides with a CH_2OH group on the carbon C-5.

The spectra of α-D-glucopyranose, ß-D-glucopyranose, α-D-mannopyranose, 1-O-CH_3-α-D-glucopyranose, irradiated in the liquid phase or in the powder state (mainly at low temperature), show two major signals (1) and (2). Their splitting constants (Table 1), retention times, stabilities are found similar for all above cited sugars.

However, these signals are not observed with ß-D-fructopyranose, ß-D-ribopyranose or α-L-rhamnopyranose (sugars with an H or a CH_3 on the carbon C-5), studied in the same conditions. This suggests that the presence of a CH_2OH group, linked to the carbon C-5 of the monosaccharide, is necessary for the radicals to be formed and trapped [2,3]. To check this hypothesis, specifically labeled sugars, [^{13}C-6]-α-D-glucopyranose, [^{13}C-1]-, [^{13}C-2]- and [^{13}C-4]-α-D-glucopyranose were experimented. The recorded spectra are all similar to that of α-D-glucopyranose, except for [^{13}C-6]-α-D-glucopyranose. In this case only, the computer analysis indicates that the signals (1)* and (2)* of the two main spin trapped radicals, both undergo a further splitting a_{13C} = 0.46 mT and a_{13C} = 0.42 mT, respectively.

The signals of the two main nitroxide sugar radicals, spin trapped in monosaccharides with a CH_2OH group linked to the carbon C-5 of glucopyranose, are perturbed by a ^{13}C on carbon C-6. Therefore, the unpaired electron must be localized in the vicinity of the carbon C-6. The abstraction of one H, or the other, from carbon C-6, followed by the formation of two spin adducts with different geometries, is proposed.

Evidence of very similar radioinduced radicals in oligomers G3 to G7.

A complete study of radicals spin trapped in G3, gamma irradiated in the liquid phase, has been carried out. Simulations of spectra of the HPLC fractions, as well as of the global spectrum, evidence five distinct radical species (1), (2), (3), (4) and (5). The results are reported in Table 2a and Figure 1.

The higher oligomers, G4 to G7, have been studied under the same experimental conditions. The recorded spectra exhibit close similarities with G3 spectra, except for an overall broadening, which increases with the molecular weight of the oligomer. The ESR hyperfine coupling constants, found for G3, were used as the input for the computer as starting guesses, together with the Gn (n = 4 to 7) experimental total spectra. The achieved simulations show that radical (1) is spin-trapped in G3 and G4 only, and that radicals (2), (3), (4) and (5) are all found from G3 to G7, with almost identical hyperfine coupling constants, except for the linewidths [12].

So, the main gamma induced radicals, spin trapped in the first glucose oligomers, are the same; as the powder ESR spectra and the main radiolytic products from irradiated starches have been found comparable to those of α-D-glucopyranose oligomers, this result is a contribution to the knowledge of starch radiolysis mechanism.

Assignment of radicals (1) and (5) radioinduced in the oligomers.

The oligomers G2 and [^{13}C-6]-G2 have been studied in the liquid phase, under the same experimental conditions, in order to get more precise informations on the structure of spin trapped radicals. The HPLC fractions of G2 are simulated with four radicals, very similar to radicals (1), (3), (4) and (5) found in G3 (Table 2b). In [^{13}C-6]-G2, radicals (3) and (4) are found unchanged. Compared to radicals (1) and (5), the radicals (1)* and (5)* exhibit identical splitting constants a_N and a_H, with a further splitting constant a_{13C} = 0.52 mT and a_{13C} = 0.70 mT, respectively.

These results show that, among the four signals of nitroxide sugar radicals (1), (3), (4) and (5) spin trapped in G2, two (from radicals (1) and (5)) are perturbed by a ^{13}C labeling on the carbon C-6. It indicates, for those radicals, the localization of the unpaired electron in the vicinity of the carbon C-6 from glucopyranose.

TABLE 2a

Radicals	ESR hsc			Sugars				
	a_N (mT)	a_H (mT)	g	G3	G4	G5	G6	G7
(1)	1.55	.63	2.00580	+	+	?	-	-
(2)	1.55	.50	2.00595	+	+	+	+	+
(3)	1.55	.36	2.00595	+	+	+	+	+
(4)	1.52	.09	2.00595	+	+	+	+	+
(5)	1.45	.02	2.00585	+	+	+	+	+

TABLE 2b

Radicals	ESR hsc			Sugars	
	a_N (mT)	a_H (mT)	a_{13C} (mT)	G2	[^{13}C-6]G2
(1)	1.54	.65	-	+	-
(1)*	1.54	.65	.52	-	+
(3)	1.53	.35	-	+	+
(4)	1.50	.09	-	+	+
(5)	1.44	.05	-	+	-
(5)*	1.44	.05	.07	-	+

Legend to TABLES 2a and 2b

ESR hsc: ESR hyperfine splitting constants: a_N, a_H, a_{13C} are expressed in mT (1mT = 10G); accuracy: ± 0.01;
g: g-factor; relative accuracy: ± 0.00005;
Gn: O-α-D-glucopyranosyl-(1->4)$_{n-1}$-α-D-glucopyranose;
(1) to (5): radicals spin trapped in unlabeled G2 to G7;
(1)*, (5)*: radicals spin trapped in ^{13}C labeled G2.
Other conventions similar to those of Table 1.

TABLE 3

ESR hsc					Sugars				
a_N (mT)	a_{H1} (mT)	a_{H2} (mT)	a_{H3} (mT)	a_{13C} (mT)	G2	S	$CH_3\alpha G$	$CD_3\alpha G$	$CH_3[^{13}C\text{-}1]\alpha G$
1.43	.24	.20	.13	-	+	+	+	+	-
1.43	.24	.20	.13	.43	-	-	-	-	+

ESR hsc: ESR hyperfine splitting constants: a_N, a_H, a_{13C} are expressed in mT (1mT = 10G); accuracy: ± 0.01;
αG: α-D-glucopyranose;
G2: maltose;
S: sucrose;
$CH_3\alpha G$: 1-O-CH_3-α-D-glucopyranose;
$CD_3\alpha G$: 1-O-CD_3-α-D-glucopyranose;
$CH_3[^{13}C\text{-}1]\alpha G$: 1-O-$CH_3$-[^{13}C-1]-α-D-glucopyranose.
Other conventions similar to those of Table 1.

Assignment of a radical spin trapped in sucrose and in maltose G2.

The spin trapping of sucrose, as well as of G2, irradiated in the powder state at room temperature, gives rise to the same preponderant radical (Table 3). A similar radical has been recorded under the same conditions with 1-O-CH_3-α-D-glucopyranose, but failed to be observed in the case of 1-O-CH_3-ß-D-glucopyranose, ß-D-glucopyranose and α-D-glucopyranose. It indicates, first, that the α-glycosidic linkage is necessary and, second, that this radical is independent of the nature of the R1 group (see Figure 2a). This strongly suggests a radical localized on the glucopyranose, probably on the carbon C-1.

To assert this hypothesis, 1-O-CD_3-α-D-glucopyranose, then 1-O-CH_3-[^{13}C-1]-α-D-glucopyranose, have been investigated under identical experimental conditions (irradiation in the powder state at room temperature) and the recorded spectra have been compared to that of 1-O-CH_3-α-D-glucopyranose. In the former case, the same radical was spin-trapped. In the latter case, the spectra analysis reveals a radical species with a further splitting a_{13C} = 0.43 mT (Table 3). These results are fully consistent with a radical on the C-1 of the glucopyranose and support the structure shown on Figure 2b. The hydrogen coupling constants a_{H1} to a_{H3} may arise from the interaction of the unpaired electron with the equatorial C-1 hydrogen, then, from two long range couplings with the axial C-3 and C-5 hydrogens [13,14].

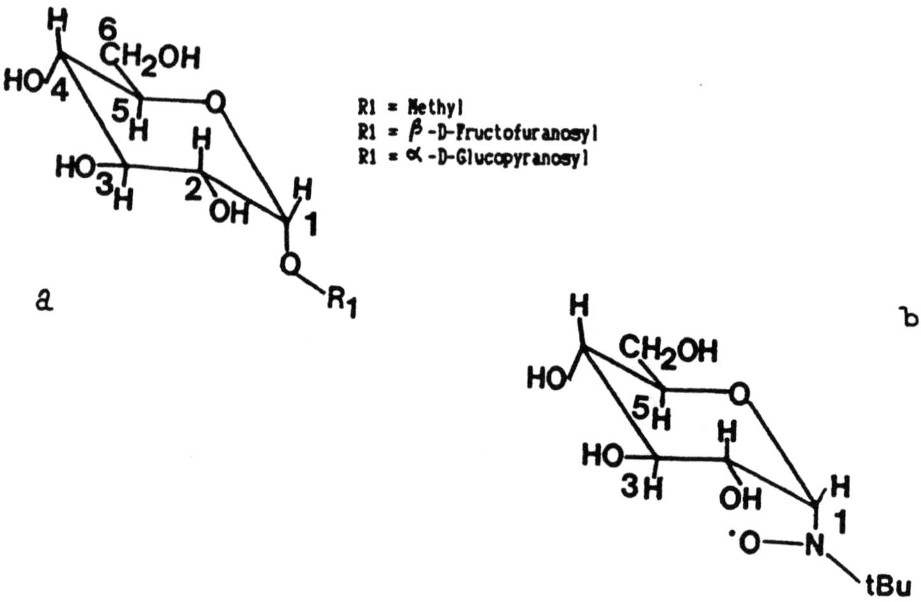

Figure 2

2a: General structure of the 1-O-glucopyranose derivatives, showing the radical with splitting constants reported on Table 3, when irradiated in the powder state at room temperature.

2b: Proposed structure for the corresponding spin trapped radical.

CONCLUSION.

This comparative study of several gamma irradiated mono- and oligosaccharides evidences:

1 - the particular gamma sensitivity of the CH_2OH group on the carbon C-5 of glucopyranose, in the case of 1-O-α-D-glucopyranose, 1-O-ß-D-glucopyranose and G2, irradiated in the liquid phase, or in the powder state, mainly at low temperature;

2 - the similarity of the five major spin adducts recorded from oligomers G3 to G7 (and of four of them from G2 to G7) irradiated in solution; this is of main interest, as glucose oligomers are good models for the study of starch radiolysis mechanism;

3 - the sensitivity of the α-glycosidic linkage, with formation of a radical localized on the carbon C-1 of the glucopyranose, in the case of 1-O-CH_3-α-D-glucopyranose, G2 and sucrose, when irradiation is carried out in the powder state at room temperature.

Synthesis and experimentation of chemically modified sugars and of other specifically labeled compounds, such as sucrose and G3, are in progress in the Laboratory. It should provide new informations on the nature of the free radicals gamma induced in sucrose and in the first α-D-glucopyranose oligomers, and, thus, on their radiolysis mechanism.

ACKNOWLEDGEMENTS

One of us (CT) is indebted to the "Institut National de la Santé et de la Recherche Médicale" for financial support. We would like to thank P. LEBEL and C.SCHNEIDER (CEN Saclay) for their specially built 8" disk drive controller (CS20), allowing the transfer of spectroscopic data from the ASPECT 2000 to IBM Compatible micro-computers.

REFERENCES.

1 Madden K (1980) Thesis, University of Rochester, Rochester, N.Y..

2 Kuwabara M, Lion Y, Riesz P (1981) E.S.R. of spin-trapped radicals from sugars. Reactions of hydroxyl radicals in aqueous solutions and gamma-radiolysis in the polycrystalline state, Int J Radiat Biol 39:4: 451-455.

3 Thiéry CJ, Agnel J-PL, Fréjaville CM, Raffi JJ (1983) Electron Spin Resonance Spin-Trapping Analysis of Gamma-Induced Radicals in Polycrystalline α-D-Glucose, J Phys Chem 87: 4485-4488.

4 Raffi JJ, Vincent PB, Agnel J-PL, Battesti CM, Thiéry CL (1987) High Performance Liquid Chromatography - Electron Spin Resonance Analysis of Sugars Irradiated in the Solid and Liquid Phase, J Chem Soc, Faraday Trans I:83:225-230.

5 Janzen EG (1971) Acc Chem Res 4:31-40.

6 Zubarev VE, Belevskii VN, Bugaenko LT (1979) The Use of Spin-Trapping in the Study of the Mechanisms of Radical Processes, Russian Chemical Reviews 48:8:729-745.

7 Suzuki N, Makino K, Moriya S, Rokushika F, Hatano H (1981) J Phys Chem 85:263-268.

8 Agnel J-P, Thiéry C, Battesti C, Vincent P, Raffi J (1985) High Performance Liquid Chromatography and Electron Spin Resonance studies of some sugar-nitroxide solutions, Anal Lett 18:A8:1013-1025.

9 Thiéry JM (1985) *VOYONS*, Programme de Simulations Conversationnelles en Physico-Chimie et en Agronomie, in: Logiciels pour la Chimie, G-M Côme, J Ducloy, E Soulié et JM Thiéry (eds) (1985) Soc Fr Chimie (Paris) et Assoc Nat Logiciel (CNRS, Nancy) 156-157.
10 Thiéry JM, Spectroscopic Applications of the *VOYONS* Program 26-30/03/1990, 23rd Annual International Conference "ESR of Radicals in Organic and Bio-Organic Systems", Egham,UK, Communication Abstract n°33.
11 Nelder JA, Mead R (1965) Computer J 7:308.
12 Thiéry JM, Thiéry CL, Agnel J-PL, Vincent PB, Battesti CM, Raffi JJ, Evans JC (1990) Electron Spin Resonance Study of Spin-Trapped Radicals from Gamma Irradiation of Glucose Oligomers, Magnetic Resonance in Chemistry 28:594-600.
13 Triolet J, Thiéry C, Agnel J-P, Battesti C, Raffi J, Vincent P (1990) E.S.R. Spin Trapping Analysis of Gamma Induced Radicals in Sucrose, Free Rad Res Comms 10:1-2:57-61.
14 Fitchett M, Gilbert BC, Willson RL (1988) Fragmentation Reactions of Radicals formed from Sugar Phosphates and the Hydroxyl Radical: an Investigation by Electron Spin Resonance Spectroscopy and Pulse Radiolysis, J Chem Soc Perkin Trans II:5:673-689.

Compatibility of Packing with Food

RIQUET A.M. (1), AKERMANN O. (1), FEIGENBAUM A. (1), PASCAL G. (1)
GAUDEMER A. (2), SANDRAY Y. (3)

(1) INRA, Laboratoire de Nutrition et Sécurité Alimentaire,
78352 Jouy-en-Josas Cedex, F
(2) Université Paris-Sud, UA 255, 91405 Orsay, F
(3) Dorlyl, Usine du Havre, BP 1152, 76063 Le Havre Cedex, F

In the food packaging sector, there is a continuous developpment of plastics and their possibilities: packaging under vacuum or under a controlled or modified atmosphere, aseptic packaging, ionization of the products in the packaging, etc...

To ensure that food manufacturers make the best possible use of these new packaging techniques, it is becoming essential to have a perfect understanding of the changes and modifications inherent in the phenomena of transfer between packaging and food.

The study which we undertook by ESR spectroscopy enabled us to acquire new knowledge about the mechanisms of interactions between container and contents and their influence on the compatibility of packaging with food:

- The behaviour of spin probes separately incorporated in rigid PVC enabled us to demonstrate a modification of the structure of the polymer after contact with conventional simulators. We were also able to underline the influence of the molecular structure of these probes on their behaviour in the material: mobility, ability to migrate.

- We also used spin labels as models for simulating the incorporation of certain constituents of the food (additives, flavourings) into the packaging. This study, which complements the previous study, gave us a better understanding of the factors which influence the absoprtion of the molecules by the polymer and enabled us to demonstrate the importance of the relative affinity of the molecules for the container and the contents.

B. Catoire (Editor)
Electron Spin Resonance (ESR) Applications
in Organic and Bioorganic Materials

I - STUDY OF THE MECHANISM OF INTERACTION BETWEEN PACKAGING AND FOOD. INFLUENCE ON THE COMPATIBILITY

This subject continues a study which we carried out earlier by ESR spectroscopy using the spin labelling technique (1). At that time, we were able to demonstrate a new aspect of the compatibility between packaging and food: analysis of the spectroscopic characteristics of a paramagnetic analogue of BHT incorporated in rigid PVC showed that the polymer structure changes after contact with aqueous media. The results obtained suggest that it is associated with the penetration of the contacting media into the macromolecular network. This result had never been demonstrated before. We therefore decided to explore it methodically (2) and discover all its consequences in terms of the compatibility of container with the contents.

We adopted the following procedure:

- to study the behaviour in rigid PVC, of spin probes of different structure and polarity, in order to establish a relationship between their movements in the material and their migration;

- to follow the change in the structure of the polymer subjected to the effect of heat and contact with fatty or aqueous media.

Experimental preparations and measurments

The spin-labelled molecules are shown in Table 1 (A), 5 (3) and 7 (1) were synthesized in the laboratory; the others are commercially available (Aldrich).

Except for the spin probes, the rigid PVC compositions correspond to those of mineral water bottles. They are dispensed in the form of 0.5 mm thick plaques prepared in the laboratory by calendering at 200°C. The labels are incoporated separately into the rigid PVC formulations, before conversion, at a rate of about 150 ppm (Laboratoires Doryl, Usine du Havre).

The rigid PVC samples, in the form of disks 1.5 cm in diameter, are immersed in glass cells (4) containing distilled water or methyl oleate which has been degassed beforehand with a stream of argon. The whole is kept at 50°C.

The ESR spectra are run before immersion and then at regular intervals of time after immersion. The spectra are recorded either at the temperature of the experiment or at room temperature.

TABLE 1. Spin labels

A. Mechanisms of interactions between containers and contents ; influence on the compatibility.

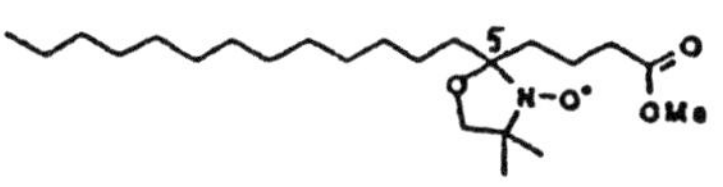

Methyl 5-doxyl-stearate (1)

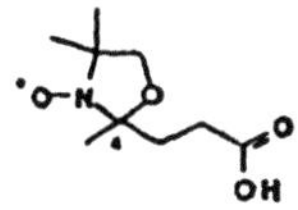

4-doxyl-pentanoic acid (4)

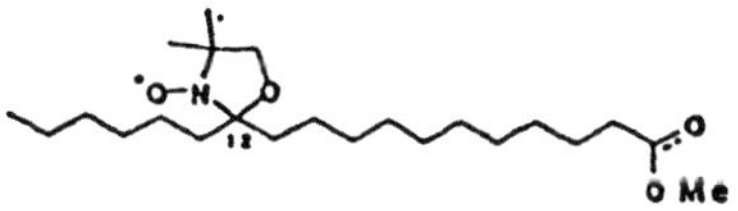

Methyl 12-doxyl-stearate (2)

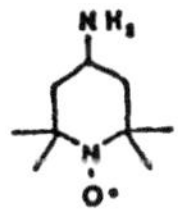

Amino-TEMPO (5)

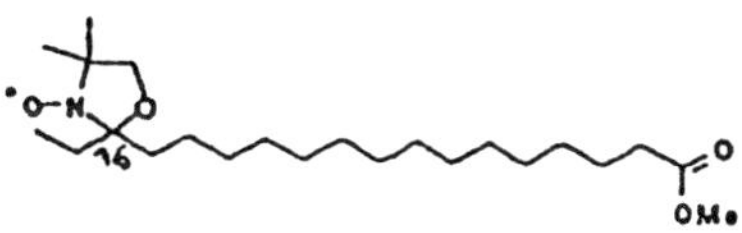

Methyl 16-doxyl-stearate (3)

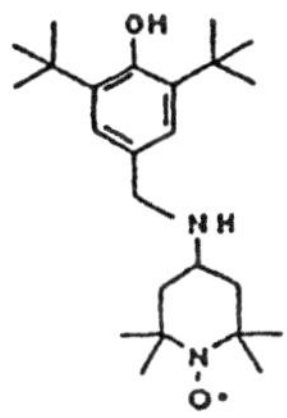

BHT-amino-TEMPO (6)

B. Study of the incorporation of the constituants of foods in crystalline polystyrene

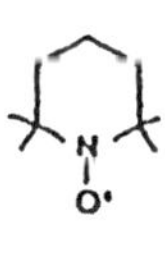

TEMPO

TEMPONE

TEMPOL

Amino-TEMPO

The respective proportions of free and immobilized labels in the material are evaluated from the composite spectra by the height of their lines at low fields (h-1/H-1 ; Figure 1).

The tensors g and T were determined by the mathematical reconstitution of experimental spectra (iterative trial-and-error method, Université Paris-Sud).

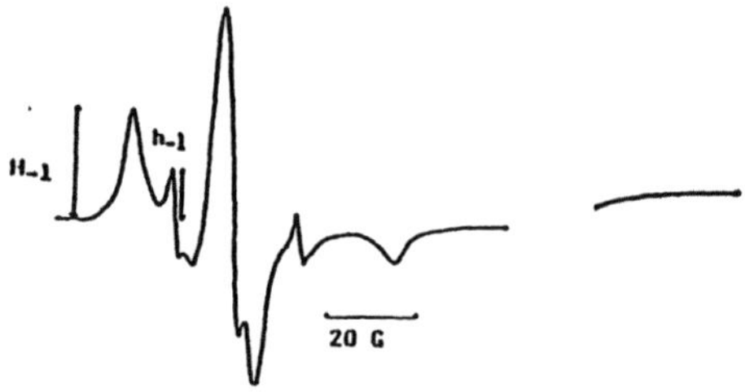

Figure 1
Measurments of ESR spectrum

RESULTS

Behaviour of the labels in the rigid PVC BEFORE contact

The analysis of successive frontal sections (10 x 30 μm) enabled us to check that the distribution of the spin labels in the PVC is homogeneous and that the shape of the spectrum of a probe on the surface and in the bulk remains the same before contact (Fig 2). This signal is characteristic of a greatly hindered nitroxide group in the polymer network; its movements are very slow (5).

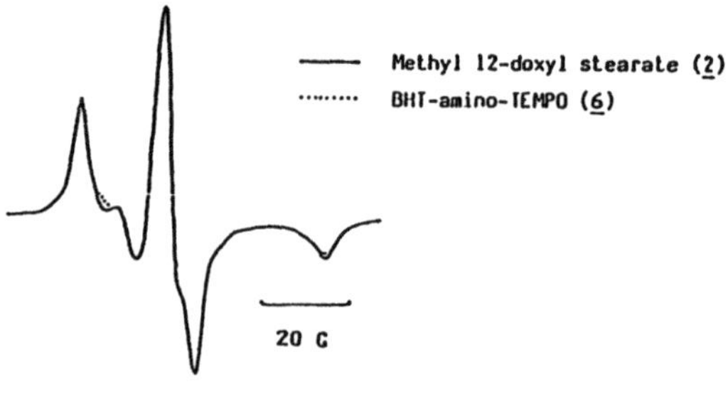

Figure 2
ESR spectrum of the PVC samples (2, 6) at room temperature

Using ESR, we were able to study the movements of the nitroxide group (NO') of the labels incorporated in the rigid PVC independently of the displacement of the molecule as a whole: the relative ease with which the NO' reorientates itself is essentially associated with the molecular structure of the label. Thus (Fig 3) the NO' group of BHT-TEMPO 5 can reorientate itself at ordinary temperature around single bonds without the molecule moving. As confirmed by the values of the orders parameter (S = 0.968), its ESR signal correponds to that of a less hindered molecule than the paramagnetic analogues of stearic acid (S = 0.985) in which the reorientation of the NO' involves the flipping of the hydrocarbon chains on either side of the doxyl.

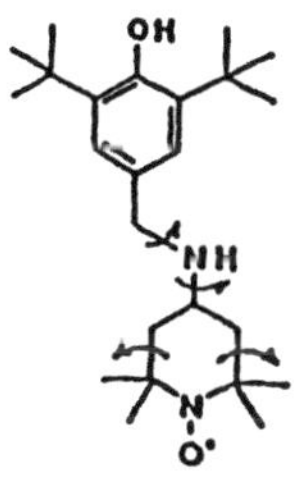

6

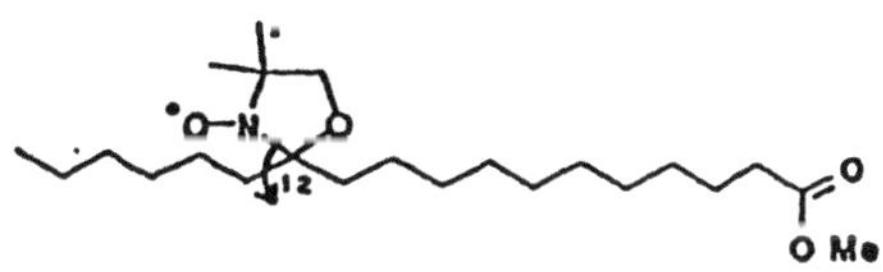

2

Figure 3
Reorientation of the nitroxide group of BHT-amino-TEMPO (6) and methyl 12-doxyl stearate (2) into rigid PVC at room temperature

On the other hand, by gradually increasing the temperature, we were able to observe that it is necessary to obtain a greater relaxation of the polymer structure for some of the molecules of 6 or 5 to become more free to move (Fig. 4).

To first approximation, it may therefore be considered that the doxyl stearates are molecules which are more capable of moving in the rigid PVC and consequently are able to migrate into the media with which it is in contact.

The particular behaviour of doxyl pentanoic acid - a molecule less free to reorientate itself without strain than its molecular structure would a priori suggest - makes it possible to underline the importance of the conformation of the molecules as regards their ability to migrate.

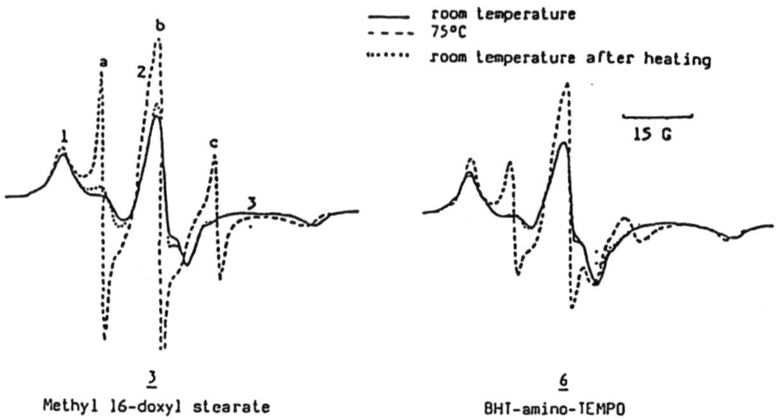

Figure 4
ESR spectrum of the PVC samples (3, 6) at different temperatures

Behaviour of the labels in the rigid PVC AFTER contact

The results obtained with 5 and 6 (Fig 5) clearly indicate that the structural modifications induced by contact with aqueous (or fatty) media are fundamentally different from those caused by a gradual increase in temperature: the spectra recorded at room temperature show the presence of a free component (lines h-1, h, h+1)which does not appear on the signal of a reference sample. Contact with water (or methyl oleate) therfore results in an increased mobility of the labels in the polymer network.

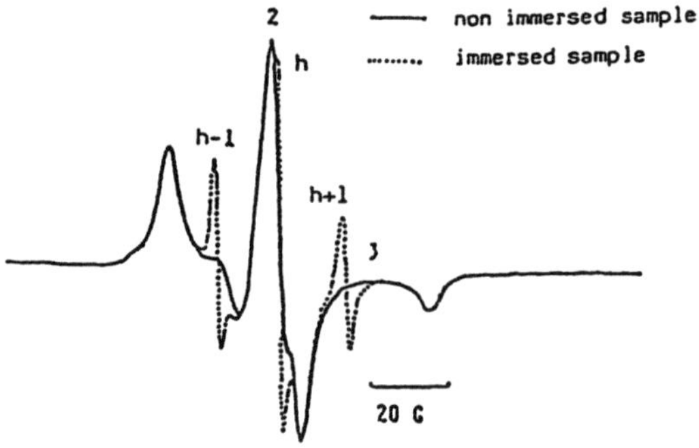

Figure 5
ESR spectrum of PVC-(BHT-amino-TEMPO) after immersio in water at 50°C
(spectra recorded at ordinary temperature)

These observations, together with those pertaining to the gain in weight of the material and its loss of transparency, strongly suggest that the structural degradation of the PVC is associated with the penetration of the media with which it is in contact. In fact, it can reasonably be acknowledged that their penetration causes certain intramolecular bonds to break (5), leading to the formation of vacant zones in which the labels are more free.

After contact, we were again able to underline the influence of the molecular structure of the probes on their behaviour in the PVC. According to their size, their shape and their polarity, they provide complementary information: where 4, 5 and 6 represent more the structural modifications of the material after contact, 1, 2 and 3 tell us about the mobility of theses probes in the PVC.

The information which can be derived from such a study ought to be able to help the plastics producers and converters in their choice of adjuvants and the modifications which have to be made in order to meet the requirements of the food manufacturers: proposal of structural criteria making it possible to design adjuvants with a low migrating capacity; use of a specific migration model.

II - STUDY OF THE INCORPORATION OF THE CONSTITUENTS OF FOODS IN CRYSTALLINE POLYSTYRENE

The work we present here forms part of a study which we undertook on the transfert of molecules from a food into a packaging (7). Its aim is to provide a better understanding of the factors which govern the absorption of certain constituents of foodstuffs (additives, flavouring etc...) by a plastic.
Electron Spin Resonance, chosen as the method of analysis, enabled us to determine, specifically and without prior extraction (8), the distribution of spin labels of variable polarity in the different phases of a food, to compare their kinetics of absorption by the material on contact with the food and to underline the importance of the relative affinity of these probes for the packaging or the food.

Experimental preparations and measurements

The spin-labelled molecules are shown in Table 1 (B). They are tetramethylpiperidinyloxy molecules: TEMPO, TEMPONE, TEMPOL and amino-TEMPO.

The chosen foods are of popular consumption and have variable fat contents: water, cream cheese (0 and 40% fat), mayonnaise (oil-in-water emulsion).

The plastic is a non-oriented crystalline polystyrene. It is in the form of 0.5 mm thick, thermoformed containers.

The spin labels are dissovled separately in the chosen food so as to give a concentration of approximately 10^{-3} M. After argon has been bubbled through, this mixture is deposited on the inner face of the polystyrene so as not to make contact with the edges (9). The whole is placed at food presservation temperature (4°C). At regular intervals of time, a 1.5 x 1 cm sample of polystyrene is taken and then analyzed by ESR after washing. The stability of the nitroxides in the food is checked under the same time and temperature conditions.

For the spectra composed of three fine equidistant lines, the number of nitroxide molecules detected is evaluated by the height of the lines (I_1, I_2 or I_3).

The levels of incorporation of the different labels were compared after two days of contact between the food and the material. This comparison is given as a relative percentage from the double integration for the spectra.

The kinetics corresponding to the absorption of TEMPO dissolved in different foods were plotted by measuring the height of the second line of the spectrum as a function of time ($I_2 = f(t)$ since the spectra are homothetic).

RESULTS

Distribution of the spin labels in the food

In 40% cream cheese or mayonnaise (oil-in-water emulsion), the spectrum of certain labels shows a splitting of the third line (Fig 6). This splitting is due to the distribution of the label between the lipid phase and the aqueous phase of the food. The line at low field (H) corresponds to the label in the aqueous phase.

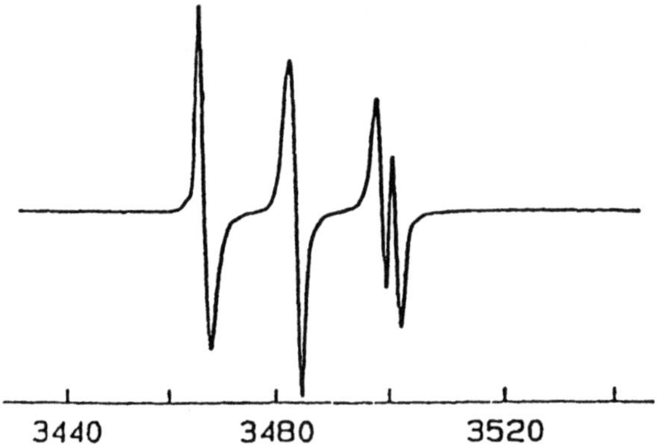

Figure 6
ESR spectrum of TEMPO in 40% cream cheese

These two lines are sufficiently fine (<1.5 G) for the coefficient of distribution of the labels in the lipid phase to be calculated according to formula (10): f = H/H+P.

The results given in Table 2 show that this coefficient depends on the one hand on the polarity of the molecules and on the other hand on the volume of the lipid phase relative to that of the aqueous phase of the food.

Shape of the ESR spectra of the labels in the crystalline polystyrene

When TEMPO, TEMPOL or TEMPONE (Fig 7) - small molecules of approximately spherical shape - are incorporated in the crystalline polystyrene after contact with the food, their signal is composed of three fine equidistant lines. It is characteristic of two signals of NO groups having a high mobility in the material.

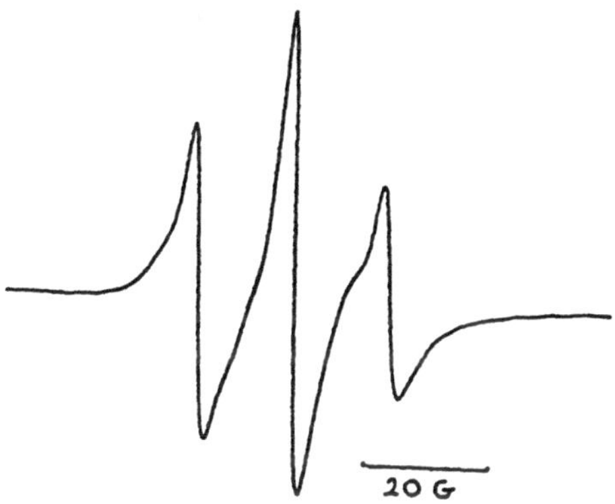

Figure 7
ESR spectrum of TEMPO incorporated in cristalline polystyrene

For amino-TEMPO, a composite spectrum is observed which corresponds to the superposition of two signals. One is characteristic of $NO^{\cdot}$ immobilized in the material and the other is characteristic of the same group having greater freedom of movement. This result is probably associated with the ionizable nature of this label.

Factors influencing the absoption of the labels by crystalline polystyrene

Influence of the polarity of the labels

Study of the incorporation of labels dissolved separately in 40% cream cheese shows (Fig 8) that TEMPO is incorporated in polystyrene about 10 times more than TEMPONE and nearly 50 times more than TEMPO or amino-TEMPO. These results can be directly related to the distribution of the labels in the lipid phase of the food (Table 2) and hence to their polarity.

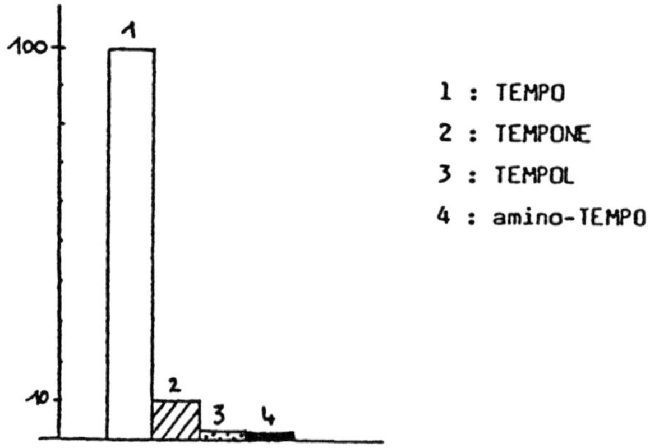

Figure 8
Influence of the polarity on the incorporation ratio of the spin labels

TABLE II Partition coeficient of the labels in the lipid phase of the food

Spin labels \ Food	cream cheese (lipid phase 20 %)	mayonnaise (lipid phase 80 %)
TEMPO	60	100
TEMPONE	10	86
TEMPOL	0	60
Amino-TEMPO	0	0

The ability of the labels - which are of similar molecular structure - to pass from the food to the crytalline polystyrene is inversely proportional to the polarity of the probe.

Influence of the nature of the contacting medium

This study was performed by comparing the levels of incorporation in the cristalline polystyrene, of TEMPO dissolved in aqueous foods: water, 0% cream cheese and 40% cream cheese and in an oil-in-water emulsion: mayonnaise.

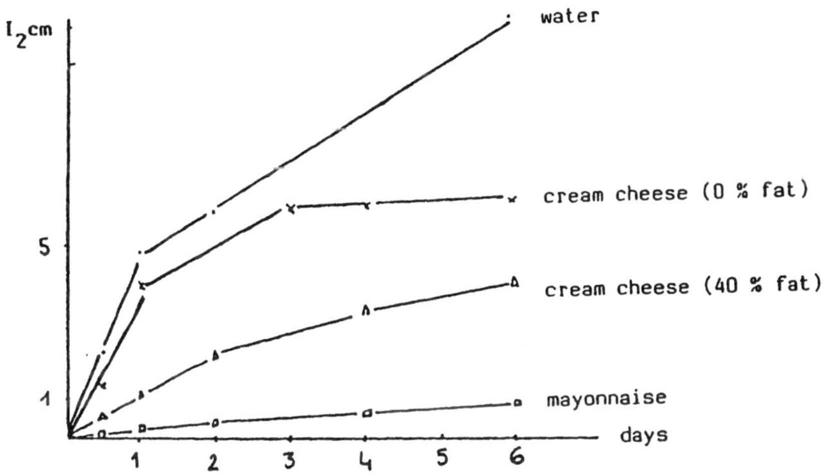

Figure 9
Kinetics of TEMPO incorporation in cristalline polystyrene from differents foods

Figure 9 clearly shows that the incorporation of TEMPO depends on its coefficient of distribution in the different phases of the food. It increases with the water content of the food.
In the case of 40% cream cheese, where the distribution of TEMPO is 60% in the lipid phase and 40% in the aqueous phase, the polystyrene/food system may be considered as a three-phase system in which the label is distributed according to its relative affinity.

The study which we undertook on the phenomena of transfer between food and packaging was carried out by Electron Spin Resonance. This methodology has the advantage that foods can be worked on without the use of conventional simulators. It enabled us to calculate the coefficient of distribution (f) of spin labels derived from TEMPO in the lipid phase of foodstuffs. We were thus able to underline the importance of the polarity of these labels on their level of incorporation in crystalline PS and the role of their relative affinity for the packaging or the food.

REFERENCES

(1) RIQUET A M. AKERMANN O. GAUDEMER A. PASCAL G (1987) Food Chem 26 271

(2) RIQUET A M. AKERMANN O SANDRAY V. FEIGENBAUM A (1991) Sci Alim 11 2

(3) JOY J. GHING-SAN (1988) J Lipid Res 29 : 1101

(4) SNYDER R C. BREDER C.V (1985) JAOAC 68 : 770

(5) BERLINER L. (1976) Spin Labelling 1 Academic Press

(6) KAMPOURIS Y REGAS F (1975) Polymer 16 . 840

(7) AKERMANN O WOLFF N RIQUET A M (1991) Sci Alim 11 1

(8) CROMPON T R (1979) Additive migration from plastics into food. Pergamon Press

(9) FIGGE K (1988) Food additives and contaminants 5 1

(10) SCHEIER S. POLNASZEK C F SMITH I C P (1978) Biochim Biophys.Acta 515 :395

Determination of the Thickness of Thin Liquid Films on Curved Surfaces

Jean-Marc di Meglio and David Quéré
UA 792 du CNRS
Laboratoire de Physique de la Matière Condensée
11, place Marcelin Berthelot
75231 PARIS CEDEX 05 FRANCE

Abstract : It is difficult to measure with accuracy the thickness of thin liquid films deposited on *highly curved* surfaces. We show how the incorporation of spin labels in the liquid can provide with a simple way for measuring such thicknesses and we present a simple illustration of this method by studying and discussing the drainage of capillary tubes.

INTRODUCTION

The study of wetting phenomena on highly curved surfaces is of crucial interest in many practical and industrial fields : let us mention for instance the adhesion in composite materials (as carbon reenforced resins), the oil recovery from porous rocks, or the coating of fibers by additives. As a consequence, methods for investigating thin films on curved surfaces (e.g. fibers, the inner of capillary tubes, of rock pores) are needed to fully characterize static or dynamical behaviors. Classical methods for determining the thicknesses of liquid films on solids, as ellipsometry or light reflectivity measurements, require perfectly smooth and flat substrates. We have thus undertaken to measure the thickness of thin films using the paramagnetic resonance of nitroxide probes incorporated in the liquid.

EXPERIMENTALS

We simply add a paramagnetic probe to the liquid ; the higher the concentration, the higher the sensitivity. As we have principally

B. Catoire (Editor)
Electron Spin Resonance (ESR) Applications
in Organic and Bioorganic Materials

studied non-polar liquids (as silicone oils), the label that we used was the common Tempo of formula :

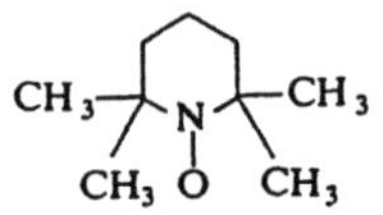

We illustrate this simple method by studying the remaining film left behind by a falling drop inside a capillary tube [1]. This study may mimic the drainage of a porous medium by gravity.

The experimental set-up is pictured in Figure 1a.

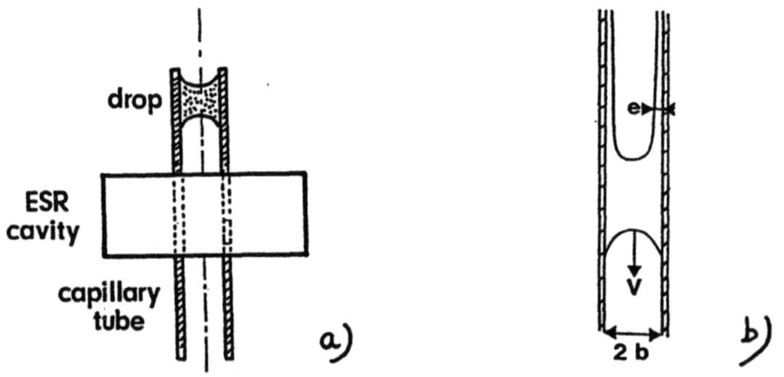

Figure 1. a) Experimental set-up : a capillary tube is set inside the ESR cavity (E9 Varian spectrometer) in place of the original Dewar. b) A drop falling inside the capillary tube and leaving a film behind.

At t = 0, a drop of labeled liquid is introduced at the top of the capillary tube (Figure 1a). The ESR signal amplitude as a function of time is reported just below on Figure 2 (as we used high concentrations of spin labels (3.10^{-2} mol/l), ESR spectra are composed of a single peak because of exchange interactions). This amplitude is the convolution product of the thickness of the film with an apparatus function f(x) which takes into account the concentration of the label inside the liquid, the settings of the spectrometer and the cavity characteristics

$$S(h,t) = \int_{-a}^{+a} f(x)\, e(h-x,t)\, dx$$

with 2a the size of the cavity (23 mm). f(x) is determined using a standard calibration procedure.

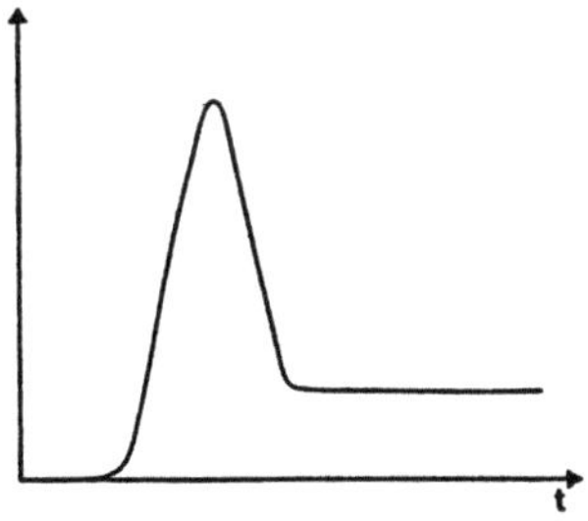

Figure 2. Schematical plot of the ESR signal (i.e. the quantity of labelled liquid passing through the microwave cavity) with respect to time. The peak corresponds to the falling drop while the plateau measures the thickness of the remaining film.

For large drops, the falling velocity V is given by

$$V = \frac{b^2}{8\eta}\rho g$$

b is the inner radius of the capillary, η the liquid viscosity, ρ the liquid density and g the gravity acceleration. The thickness of the film left by a moving interface in a capillary tube was derived by Deryaguin [2] and is given by

$$e = 1.3\, b \left(\frac{\eta V}{\gamma}\right)^{2/3}$$

with γ the surface tension of the liquid. Combining these two formula gives (by elimination of V) :

$$\frac{e}{b} = 0.325 \left(\frac{b}{\kappa^{-1}}\right)^{4/3}$$

where we have introduced the capillary length $\kappa^{-1} = (\gamma / \rho g)^{1/2}$. It is remarkable that this thickness does not depend on the liquid viscosity.

The following figures (Figures 3a and 3b) represent our experimental results along with the theoretical predictions which are found to correctly describe the findings.

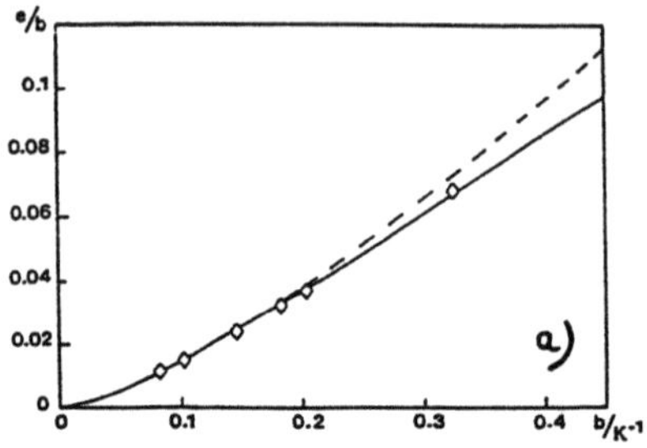

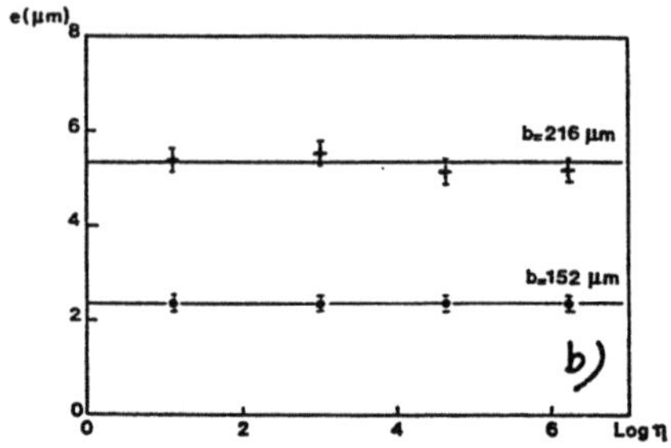

Figure 3. a) Thickness of the remaining film (divided by the capillary radius b) after the drop has passed versus b/κ^{-1}. Diamonds are experimental points. The dotted line represents the theory developed in the text and the full line is a modification of this theory which takes into account film thickness and gravity effects [1].
b) Thickness of the remaining film for two different radii of the capillary tube and four viscosities (η = 3, 20, 100, 500 cSt). Full lines are theoretical predictions.

Because of the great sensitivity of ESR, it is then possible to investigate the drainage of the film for long periods of the order of one day (during the drainage process, drops are forming at the bottom of the capillary tube, imposing a suction of magnitude $-\gamma/b$; this capillary suction is more efficient that gravity for drainage). The following picture (Figure 4) shows on a log-log plot the decay of the thickness as a function of time.

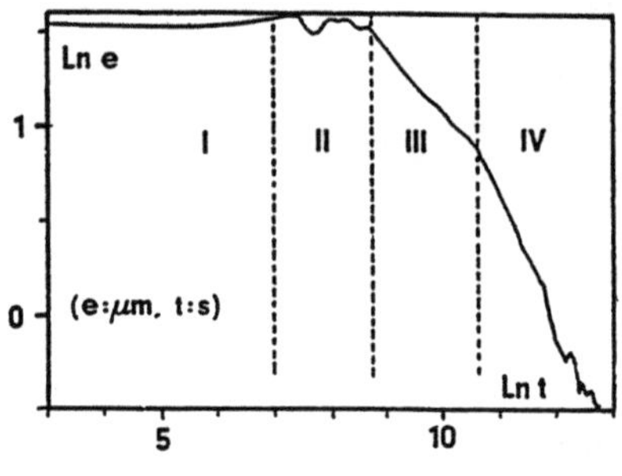

Figure 4. Thickness of the film versus time. The capillary radius b is 0,216 mm and the silicone oil viscosity is 500 cP.

We can distinguish four regions of interest :

- region I : the thickness is essentially constant and is equal to the thickness just after the falling of the drop (see figure 3).

- region II : we notice oscillations of the thickness that correspond to drops crossing the cavity. These drops are due to the onset of the well-known Rayleigh instability : any liquid column or annular column is unstable because of the surface tension. Because of the downward flow, this instability then "saturates" (this can be explained by incorporating non-linear effects in the Stokes equation [3]) : the oscillations disappear.

- regions III and IV : well defined power-law dependences ($e \sim t^{\alpha}$) are observed. We get respectively $\alpha = -0.3$ and $\alpha = -0.67$ for regions III and IV. The drainage rate in these regions should be described using a simple lubrication approximation since the implied thicknesses are small, i.e. :

$$\frac{\partial e}{\partial t} = \frac{1}{3\eta} \frac{\partial}{\partial h}\left(e^{3} \frac{\partial P}{\partial h}\right)$$

with P the pressure inside the film. The power law of thickness with time depends on the ingredients that we put inside this pressure : pure gravity ($P = \rho g h$) yields $\alpha = -1/2$, Laplace pressure due to curvature ($P = \gamma/(b-e)$) yields $\alpha = -1/3$. We can then say that the drainage process in region III is due to the remaining drop remaining at the bottom of the tube (which is reasonable since the cavity is located at a few centimeters above the lower extremity of the capillary tube : then $\rho g h < \kappa^{-2}/b$). The faster regime occuring in region IV may be due to the presence of axial grooves on the inner surface of the capillary tube [4] (an additional suction by grooves of depth of order 1 micron can explain the observed behavior).

The drainage can then be followed by ESR at even longer scales corresponding to smaller thicknesses than represented here in order to investigate the role of long range forces (as van der Waals) [5], but it should be emphasized that these forces can be changed by the presence of labels and by the eventual adsorption of these labels on the tube surface [6].

CONCLUSION

We have shown on a simple example that the use of the spin labeling technique can provide with a powerful tool to investigate both the statics and the hydrodynamics of thin liquid films. This technique is particularly well adapted for our problems, in which the solids coated with liquid are highly curved.

1 D. Quéré, J.-M. di Meglio (1989) C. R. Acad. Sci. Paris, Série II, 308:147
2 B.V. Deryaguin (1945) Acta Physicochim. USSR 20:349
3 D. Quéré (1990) Europhysics Letters, in press
4 D. Quéré (1989) Thèse de Doctorat, Université de Paris VI
5 D. Quéré, J.-M. di Meglio, F. Brochard-Wyart (1990) Science, 249:1256
6 D. Quéré, J.-M. di Meglio, F. Brochard-Wyart (1988) Rev. Phys. Appl. 23:1023

Electron Spin Resonance Measurement of Nitroxy Probes Mobility, Attached to a Polymer Through a Spacer Arm

A. GUYOT [1], A. REVILLON [1], M. CAMPS [2], JP. MONTHEARD [2] and B. CATOIRE [3]

1 CNRS - LMO BP 24 F-69390 Lyon Vernaison, France
2 Université de Saint Etienne, 23 Rue du Docteur Michelon, F-42100 St Etienne France
3 Insititut Textile de France, BP 60, F-69132 Ecully Cédex, France

SUMMARY

Stable nitroxide radical probes are attached to styrene-divinylbenzene gel type resins through spacer arms, the length of which being 1 to 7 methylene groups. Results of electron spin resonance analysis which allows to estimate the local viscosity are discussed. Mobility of the nitroxide probes generally increases with the number of CH_2 groups, the effect being compared to a decrease of cross-linking density.

INTRODUCTION

It is now well established that a functional group attached to a polymer through a spacer is more active than its homologous directly fixed to the polymer (1, 2). One of the proposed explanations is the decrease of steric hindrance when the site is at a greater distance from the surface. This may be true when the support is an inert, non-porous, solid, such as silica. The situation is quite different with a polymeric support. The best representation of a swollen crosslinked polymer bead is to consider it as a semi-dilute solution of polymer segments with a variable local viscosity, sometimes very high. The spacer may increase the local mobility of the functional group in the medium, but its mobility remains lower than that of the corresponding unattached molecule. Our approach was to use the spin-labelling technique to evaluate the functional site mobility, in the swollen state, when a stable free radical NO· is attached to it. The rotational correlation time τ, determined from the ESR spectrum is porportional to the internal viscosity. Regen has found that τ decreases when swelling increases, depending on the nature of solvent or when crosslinking density decreases,

B. Catoire (Editor)
Electron Spin Resonance (ESR) Applications in Organic and Bioorganic Materials

according to divinylbenzene amount, in the range 2 to 12%, in relation with the mobility of substrates bound to the gel (3, 4, 5). The purpose of this work is to determine the effect of different length of spacers attached to support of different crosslink density. Moreover the studied samples have a morphology different from Regen's ones.

EXPERIMENTAL

Preparation of monomers

Monomers 2a and 2b

The Grignard compound of vinylbenzylchloride 1 (VBC, commercial mixture 60/40 meta/para) is prepared by a slow addition of 15.2 g (O.1M) of VBC in 300 mL of anhydrous ether to 2.5 g of magnesium powder stirred with 20 mL of diethyl ether. After completion of reaction, the Grignard reagent is added to a solution of 21.6 g of 1,4-dibromobutane (0.1 M) -for 2a and 24.4 g of 1,4-dibromohexane (0.1 M) for 2b- in 100 mL of tetrahydrofuran with Immol of lithium tetrachlorocuprate according to the procedure described by Tomoi et al (2,6). The reaction is allowed to proceed 48 hours instead of 5. The yields are closed to 40%. Monomer 3.

Ethylene oxide is reacted with 4-chloromagnesium stryrene, then the 4-(2-hydroxyethyl) styrene is treated by triphenylphosphine and carbon tetrachloride, as described in (6).

Physical constants (IR and NMR) are in agreement with those described in (7).

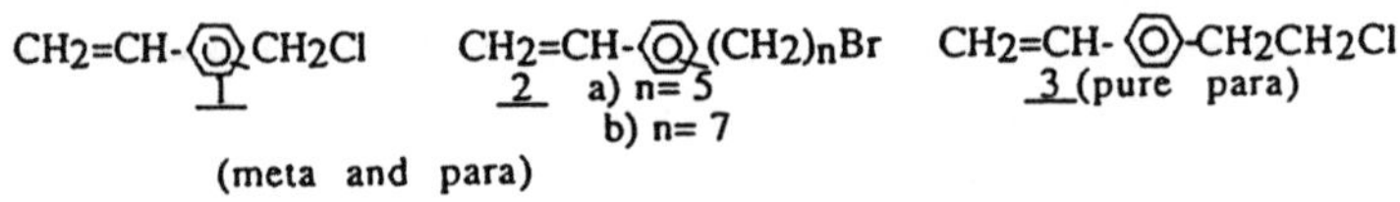

Synthesis of polymers

The styrene (St) is a commercial one (Prolabo, 99% purity, 30 ppm t-butyl-4-catechol stabilizer). Divinylbezene (DVB) is the conventional mixture of 55% pure DVB (isomers %: 71 meta, 29 para) and 45% ethylvinylbenzene (EVB) in presence of ≈ 600 ppm 2,6-t-butyl-4-methylphenol.

Free-radical copolymerization in suspension is performed at 88°C, under nitrogen, for 8hours, with benozyl peroxide as initiator (1% weight of monomers). The organic phase is diluted (1:1 in volume) by toluene as a solvent porogen. The volume ratio of aqueous phase to organic phase is 5. Stirring rate is 230 rpm. The dispersive agent is the ammonium salt of an alternate styrene-maleic anhydride copolymer (Lytron) in solution 2.5% in water. The protective colloid is an arabic gum, 2% weight in deionised water.

After reaction, the beads are washed with water, then methanol, extracted with methanol in a Soxhlet (≈ 10 cycles) and finally dried at 60°C for 24 hours under vacuum.

A standard recipe is indicated in Table 1.

Table 1 : Syntehsis of crosslinked terpolymers

Organic phase					Aqueous phase		
Styrene	DVB	VBC	Toluene	BP	H_2O	Lytron	Gum
81g	12 g	7g	110 mL	1 g	1 L	25 mL	20 g

Compositions of two series of copolymers are reported in Tables 2 and 3: respectively, constant amount of functional monomer, but varying crosslink density and at constant composition, but varying the number n of methylene groups of spacer from 1 to 7.

Spin labelling

The stable free-radical , covalently bound to the support and used as an ESR spin-label is: 4-hydroxy TEMPO or 4-hydroxy 2,2,6,6 tetramethyl piperidinolyoxy,

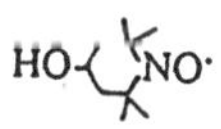

a commercial product from Aldrich. Sodium hydride is a commercial product of Merck. Dimethylformamide, from Prolabo is distillated before use on CaH_2 under N_2.

6 mg (0.035 mmol) 4-hydroxy TEMPO, 45 mg (1.9 mmol) NaH and 8 mL DMF are stirred under nitrogen at room temperature, for 4 hours. Then, this mixture is added to 2.3 g (≈ 1 mmol) of support in 75 mL of DMF, for reaction at room temperature, under nitrogen for 24 hours. The support is filtered, successively washed with water, ethanol, benzene, diethyl oxide and dried at 60°C, for 24 hours under vacuum (0.2 tor.).

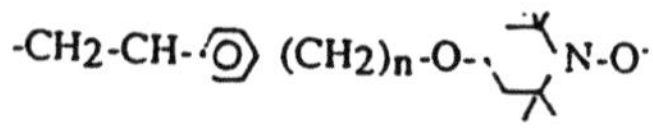

ESR measurements

- Material: The spin-labelled nitroxide polymer powder is swollen in toluene for 24 hours at room temperature and then, without drying, introduced in a quartz tube.

- Equipment: ESR spectra are obtained at 20°C, with a Bruker 100D equipped with a 10,000 gauss and working in X band.

<u>Results and discussion</u>

Figure 1 shows the spectra of the polymer including spin-labelled spacers. One can see a well resolved triplet line of 30 gauss total width.

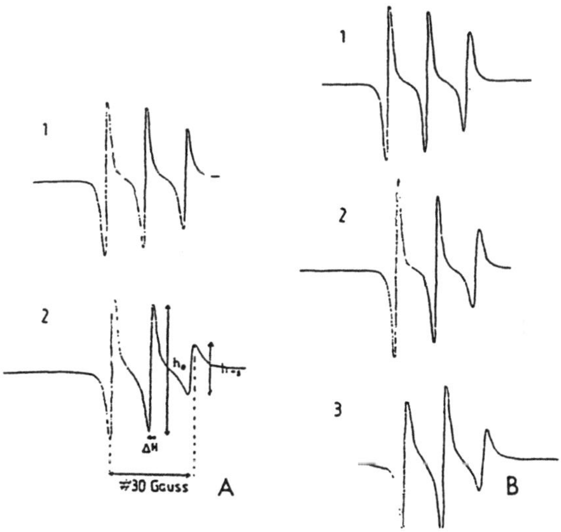

Fig 1: Room temperature ESR spectra of spin-labelled nitroide polymers, swollen in toluene: A, as a function of DVB amount 1: 1%; 2: 10% and B, versus the length of the spacer arm: 1: 7; 2: 5 and 3: 1

However the heights of the high field peaks (-1) are sensitive to the molecular structure because of the weak anisotropy of spectra due to some difficulty of the spin labelled spacer to reorientate (10^{-9} - 10^{-10} s), the correlation time is given in this case by

$$\tau\,(s) = 6.5\ 10^{-10}\ \Delta H_0\{(h_0/h_{-1})^{0.5} - 1\}$$

Good reproducibility of τ measurements for a series of a given sample has been obtained. Moreover it has been checked that the toluene solution after contacting the polymer and separation from it, does not show any significant signal. It may be concluded that the probe is well attached to the polymer and that the free spin-label is not included in the structure.

Table 2: Composition and correlations time for VBC terpolymers

Styrene	93	91	85	75
EVB	1	2	5	10
pure DVB	1	2	5	10
VBC	5	5	5	5
$10^{10}\,\tau$ (s)	3	4	9	9.5

Table 3: Composition and correlation time for functional spacer terpolymers

Styrene	85	85	85
EVB	5	5	5
pure DVB	5	5	5
Spacer	5	5	5
n and X	1,Cl	5,Br	7,Br
$10^{10}\,\tau$ (s)	9	6.3	3

The correlation time is proportional to the inner viscosity of the medium. Table 2 shows that for a series where the spacer is short (one methylene group, corresponding to vinylbenzylchloride as starting monomer), this time increases with the amount of crosslinker as already observed by Regen (3). For a series with the same crosslinking density but increasing length n of the spacer arm, the correlation time decreases, as it may be seen in Table 3. This variation is as expected. For this series the effect on τ for $n = 7$ is about the same than a decrease in the amount of crosslinker form 5 to 1%, in the other series. However, the local viscosity has been found very low: $\tau = 0.8\ 10^{-10}$s for the sample where $n = 2$. In that case, the amount of probe attached to the gel was much lower than in the other cases. Two hypotheses may be given: i) the nucleophilic reaction of the sodium salt of the probe may cause the elimination reaction instead of the expected substitution reaction

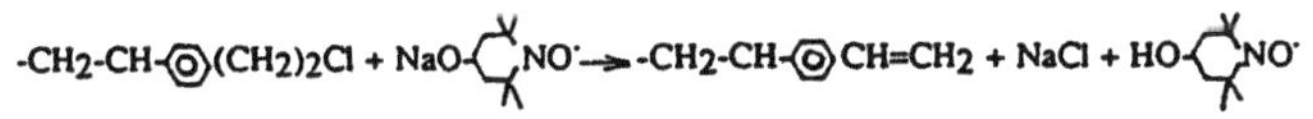

ii) the spacer arm is ended by a chlorine atom, far less reactive than that of the VBC or than the bromine atom (8) used for longer spacer arms. In all cases, the average extent of derivatization is low (3%), in order to avoid interference effects between NO˙ groups. It has been found that Br is homogeneously distributed in terpolymers (9), but from our knowledge of the elaboration of morphology, we may assume that a distribution of accessibility of sites exists. Only the more accesible sites by TEMPO give rise to the ESR signal. This may explain that the low value of τ (for $n = 2$) does not correspond to the mean internal viscosity, but to superficial mobility of some functional groups. Two rates of rotational diffusion of the spin labels have been assumed in (10), depending on their attachment on chains on the inner surface or inside zones of higher crossliking density, for macroporous copolymers. We are going on investigation in this way.

REFERENCES

(1) Tomoi M., Ogawa E., Hosokawa Y., Kakiuchi H., J. Polym. Sci., Polym. Chem. Ed 20, 3421 (1982)

(2) Tomoi M., Ogawa E., Hosokawa, Y. Kakiuchi, H., J. Polym. Sci., Polym. Chem. Ed 20, 3015 (1982)

(3) Regen S.L., Macromolecules, 8, 689 (1975)

(4) Regen S.L., J. Am. Chem. Soc. 96:16 5275 (1974)

(5) Regen S.L., J. Am. Chem. Soc. 97:11, 3018 (1975)

(6) Boinon B., Benayad B., Monthéard J.P., Makromol. Chem., 190, 241 (1989)

(7) Braun D., Keppler H.G., Makromol., chem. 78, 100 (1964)

(8) Desmurs J.R., Jouve I., Gérard B., Zimmermann P., L'Actualité Chimique, 121 (1989)

(9) Bacquet M., Lemaguer D., Caze C., Eur. Polym. J. 24, 533 (1988)

(10) Pilar J., Horak D., Labsky J., Svec F., Polymer 29, 500 (1988)

Published in "Polymer Bulletin 23 (1990)" - Springer Verlag 1990

An ENDOR Study of Methyl Dynamics at Low Temperature

F. Bonon, M. Brustolon, A.L. Maniero and U. Segre
Dipartimento di Chimica Fisica
Università di Padova (Italy)

Abstract

We have studied the motion of the methyl group in the free radical obtained by γ-irradiation of 4-methyl-2,6-di-t-butylphenol. The ENDOR spectra have been measured in the temperature range 6 to 200 K. The height of the hindering potential barrier has been obtained from the shift of the ENDOR lines at low temperature.

Introduction

Many macroscopic quantities of molecular crystals, as specific heat or dielectric constant, are related to the internal dynamics of the molecules composing the crystal. According to Press [1], the internal motions in these solids may be classified as *collective* and *single-particle* motions. Single-particle motions describe the rotation of a single molecule or a functional molecular group in its surroundings. The effect of the surrounding can be approximated as an angle-dependent potential, with a time-independent and a fluctuating part. At relatively high temperatures the fluctuating part of the potential is very important and the rotational motion is a classical jumping between the potential wells [2].At very low temperature, instead, the rotation is better described by a quantum tunnelling motion below the time-independent potential barriers.

Different techniques can be employed to study internal molecular rotations in molecular solids. Among them, magnetic resonance spectroscopies have a prominent role. In particular, ENDOR offers the unique merit that it can be used on a large temperature range, allowing the determination of the rotational potential for both the classical and quantum regimes of motion.

B. Catoire (Editor)
Electron Spin Resonance (ESR) Applications
in Organic and Bioorganic Materials

Previously we have studied the internal motions by ENDOR at relatively high temperature, in the range 100—200 K. We have shown that it is possible to obtain the value of the potential barrier in the classical regime of motion from the amplitude of the ENDOR spectra [2].

In this paper the quantum rotational motion of a methyl group is considered. In this case an accurate analysis (up to the second order in the hyperfine couplings) of the line positions of the ENDOR spectrum allows to obtain the required information on the hindering potential of the methyl group.

We have studied the the free radical **I** obtained by γ–irradiation of 4–methyl–2,6–di-t-butylphenol:

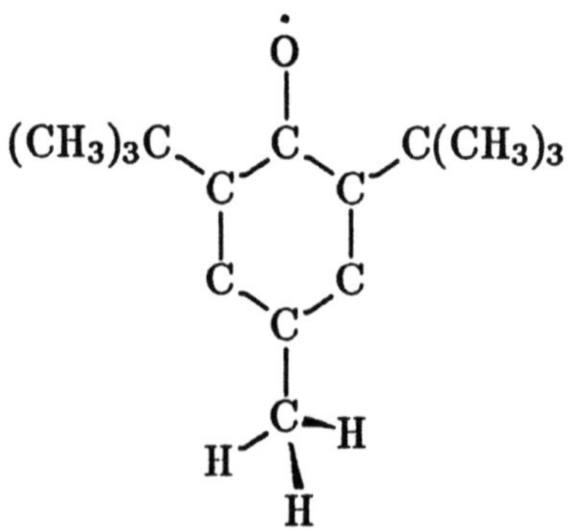

Theoretical framework

The motion of the CH_3 group is described by a single angular variable ϕ. The hindering potential of methyl groups in molecular crystal has a threefold symmetry axis:

$$V(\phi) = \frac{1}{2}V_o(\cos 3\phi - 1) \tag{1}$$

For an infinitely high barrier V_o the energy levels $\mathcal{E}_i$ are three-fold degenerate and only the librational motion is possible. For finite values of V_o the librational levels are splitted, and the methyl group undergoes a coherent rotational tunnelling motion [1]. We have a couple of degenerate states of symmetry E and a non–degenerate state of symmetry A. The ordering of these states is alternate for the different librational levels, and the tunnelling splitting values for the first two librational levels are given by:

$$\begin{aligned} \Delta_0 &= \mathcal{E}_0^E - \mathcal{E}_0^A \\ \Delta_1 &= \mathcal{E}_1^A - \mathcal{E}_1^E \\ 0 &< \Delta_0 \ll \Delta_1 \end{aligned} \tag{2}$$

The spin states must be eigenstates of the total nuclear spin **I**. The splitting between the A and E spin functions can be included in the spin hamiltonian by adding to it the spin exchange operator [3]:

$$H_t^{(i)} = -2J_i \sum_{j<k} \mathbf{I_j} \cdot \mathbf{I_k} \tag{3}$$

with $J_i = \Delta_i/3h$. The effective spin hamiltonian for the methyl group in the i-th librational level turns out to be:

$$H_{\text{eff}}^{(i)} = H_Z + H_{hf} + H_d + H_t^{(i)} \tag{4}$$

where H_Z, H_{hf} and H_d are the Zeeman, hyperfine and nucleus–nucleus dipolar interactions respectively.

At low temperature (about 6 K) only the ground librational state is occupied, and the spectra must be function of ground state splitting J_o only. By increasing the temperature, both the ground and excited librational states become populated. It has been attempted [4] to simulate the population dynamics as a classical exchange process between two sites corresponding to the librational levels with $i = 0$ and $i = 1$.

Experimental results

EPR spectra. In Figure (1.a) the EPR spectrum at low temperature (quantum tunnelling motion) is compared to that at high temperature (classical hopping motion). The high temperature spectrum has the usual 1:3:3:1 pattern which is is due to the hyperfine coupling of the electron spin with three equivalent protons, with neglect of the last term in the spin hamiltonian (4). The low temperature spectrum instead is well accounted for by the first order eigenvalues of the hamiltonian $H_{\text{eff}}^{(0)}$. When $|H_t^{(0)}| \gg |H_{hf}|$ the lineshape switchs from the classical triplet to a septet with 1:1:1:2:1:1:1 ratios [5]. Therefore, no exact information on the value of the tunnelling can be acquired from the EPR spectrum.

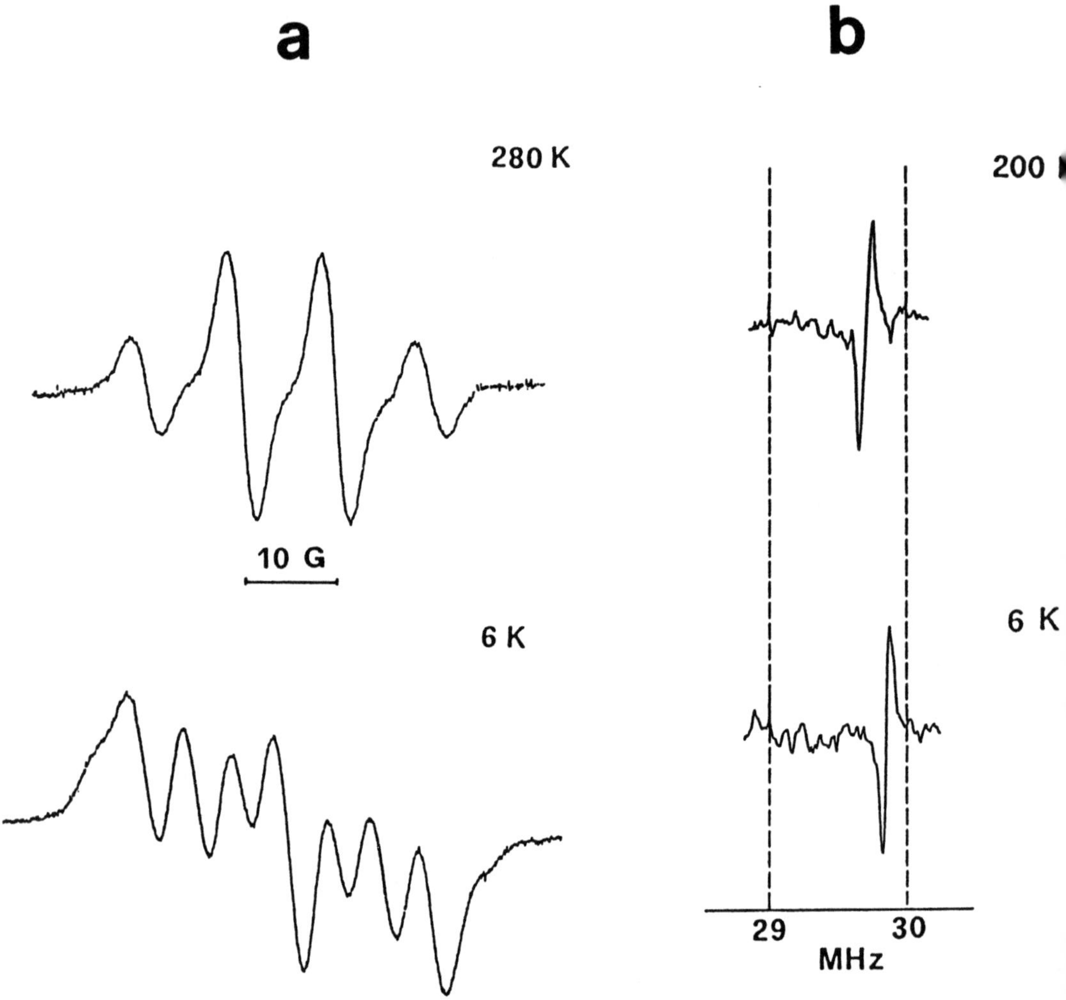

Figure 1: EPR spectrum (a) and ν_+ line of the ENDOR spectrum (b) of radical **I** in the quantum (low T) and in the classical (high T) regimes of motion.

ENDOR spectra. The crystal structure is orthorombic with four non-equivalent sites in the unit cell. The principal values and cosine directors of the hyperfine tensor for the four sites have been determined previously by measuring the angular dependence of the ENDOR ν_+ line of the methyl proton at room temperature [6].

ENDOR spectra have been recorded with the magnetic field along one of the crystallographic axes in the temperature range 6—200 K. A shift in the line position has been observed by increasing the temperature, see Figure (1.b). The line position at 6 K can be computed by solving the hamiltonian for the ground state $H_{\text{eff}}^{(0)}$ up to the second order. The fitting value of J_o is 2.1 GHz, corresponding to a hindering barrier height $V_o/k = 270$ K. To interpret the temperature dependence of the line shift, it is possible to employ the Allen's model [4].

References

1. W. Press, *Single Particle Rotation in Molecular Crystals*, Springer (1981).
2. M. Brustolon and U. Segre, in *Advanced EPR. Applications in Biology and Biochemistry.*, A.J. Hoff ed., Elsevier (1989).
3. S. Clough and F. Poldy, *J. Chem. Phys.*, **51**, 2076 (1969).
4. P.S. Allen, *J.Phys. Solid State*, C7, L22 (1974).
5. J.H. Freed, *J. Chem. Phys.*, **43**, 1710 (1965).
6. F. Bonon, M. Brustolon, A.L. Maniero and U. Segre, to be published.

Etude par Résonance Paramagnétique Electronique (RPE) de la Décroissance Radicalaire dans les Médicaments Irradiés

F.ZEEGERS et B.TILQUIN

Laboratoire d'analyse chimique et physico-chimique des médicaments, CHAM 72.30, 72, avenue E.Mounier, 1200 Bruxelles, Belgique.

RESUME

Divers médicaments ont subi l'action ionisante des rayonnements gamma du Cobalt 60 : le chloramphénicol, le diazepam et le clonazépam.

Nous nous attarderons sur le cas du diazépam pour définir dans quelle optique la RPE est utilisée pour l'analyse de médicaments irradiés. La décroissance radicalaire peut être utilisée pour définir la localisation de l'énergie cédée.

INTRODUCTION

Pour les rayonnements ionisants comme les rayons γ du Cobalt 60, l'interaction principale avec la matière est due à l'effet Compton : le rayonnement est dévié et un électron accéléré est produit in situ. Cet électron "Compton" matérialise son parcours sinueux par la formation d'espèces réactives qui ne sont pas en équilibre thermique : des radicaux-cations, des molécules excitées et des électrons secondaires.
Les radicaux libres sont formés par des réactions immédiates lors de la relaxation des espèces primaires excitées ou par des réactions ions-molécules très rapides (10^{-14} seconde).

Sur le schéma I sont résumées les principales étapes conduisant à la formation de produits finals.

B. Catoire (Editor)
Electron Spin Resonance (ESR) Applications in Organic and Bioorganic Materials

Schéma I : Principales étapes conduisant aux produits finals

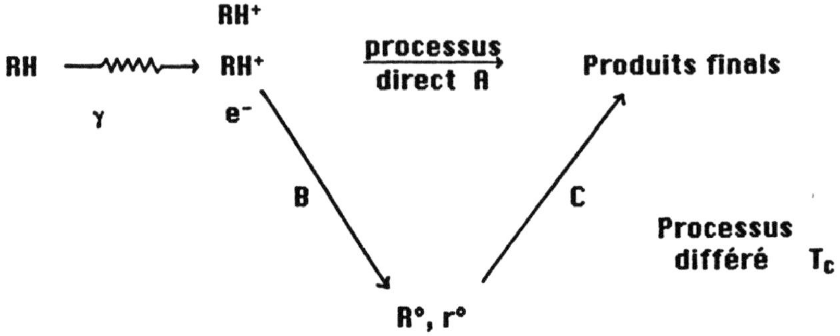

En phase solide (1), tout schéma de radiolyse doit envisager le piégeage, dans la matrice du solide, des radicaux libres radio-initiés, en particulier ceux qui sont formés par la rupture d'une liaison C-H (radicaux parents). Ce sont ces radicaux qui seront analysés en RPE. Certains radicaux fragments r° (rupture d'une liaison C-C) sont toutefois formés. La concentration relative de ces radicaux est le plus souvent très faible en phase solide à cause de "l'effet cage" : en effet, ces radicaux fragments, trop gros pour diffuser, sont contraints de rester sur leur site de formation et se recombinent en cours de radiolyse. Les radicaux fragments isolés détectés par mesure d'interception sont dus à ces processus secondaires.

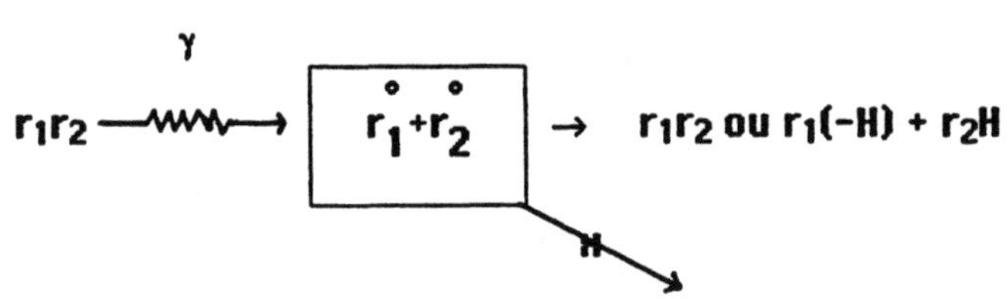

A la température T_{irr}, pendant l'irradiation du solide, se forment les produits finals de la radiolyse (voie A) et les radicaux piégés (voie B). Ces derniers réagissent entre eux à une température T_r lors du réchauffement de l'échantillon (voie C). Cette température T_r de recombinaison radicalaire est dictée par les propriétés physiques du solide irradié et non par la température d'irradiation ou la nature des radicaux. La diminution de la concentration radicalaire dans le milieu est observée par la décroissance du signal RPE (2). Lors du réchauffement de l'échantillon, les radicaux parents et fragments peuvent entreprendre différentes réactions :

1. Réaction de recombinaison ou de disproportionnement :

(1) R° + R° → RR ou R' + R" (combinaison ou - dismutation)
(2) R° + r° → Rr ou R' + r' (réaction croisée)
(3) r° + r° → r' + r" ou RH (effet cage)

2. Réaction d'arrachement d'un atome H à la molécule parente :

(4) R° + R'H → R°' + RH où R°' est le radical le plus stabilisé
(5) r° + R'H → R°' + rH

Les réactions (4) et (5) de metathèse devraient s'accompagner d'une variation lente (à 25°C) de la nature des radicaux piégés pour évoluer vers le radical parent "le plus stabilisé". Or cette variation ne s'observe pas dans l'allure du spectre RPE.

L'évolution isotherme à 25°C est négligeable pendant plusieurs mois.

De plus une réaction de metathèse efficace permettrait un déplacement du centre radicalaire avec comme finalité la rencontre et la disparition isotherme des radicaux.

3. Diffusion dans le milieu où se produit une compétition entre les réactions de recombinaison (1 à 3) et les réactions d'abstraction d'hydrogène par les radicaux (4 et 5).

En phase solide, la diffusion est limitée. Une réorganisation du solide permet la diffusion des radicaux résiduels à la température de recombinaison T_{rec}.

En phase solide, la diffusion est limitée. Une réorganisation du solide permet la diffusion des radicaux résiduels à la température de recombinaison T_{rec}.

Notons pour terminer cette introduction que l'atome H° piégé n'est pas observé par RPE à 77 K dans des solides moléculaires irradiés.

1. ASPECT EXPERIMENTAL

1.a. Le composé étudié : le diazépam (DIAZ)

Le diazépam appartient à la famille des benzodiazépines qui sont des sédatifs et des tranquillisants. Son aspect est celui d'une poudre blanche qui fond à 131.135°C. Quasi insoluble dans l'eau, il est soluble dans l'éthanol et le chloroforme.

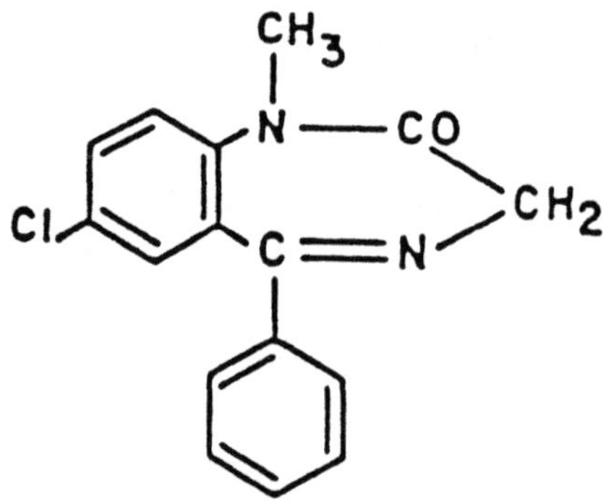

PM = 284,76

La constante d'acidité pK_a vaut 3,3 à 20°C (Merck Index).

1.b. La radiolyse

Le solide est introduit dans une ampoule sphérique d'un diamètre interne de 2 cm, fixé sur une rampe à vide. Le solide subit ainsi un cycle de dégazage. Lorsqu'il subsiste au-dessus de l'échantillon une pression résiduelle inférieure à 10^{-4} Torr, l'ampoule est scellée sous vide. L'échantillon est irradié dans la chambre panoramique de Louvain-La-Neuve au Co^{60}.

La dose est exprimée en Gray. Le Gy équivaut à l'absorption d'une énergie de 1 joule par un kilogramme de matière irradiée. On définit également le rendement radiochimique G comme le nombre d'atomes ou de molécules produits par absorption de 100 eV.

Il existe des mesures indirectes de la dose absorbée dont la plus reproductible est le dosimètre de Fricke. Il consiste en une solution aqueuse saturée en air contenant 10^{-3} M de sulfate d'ammonium de fer (II), 10^{-3} M de chlorure de sodium et 0,4 M d'acide sulfurique. Le rendement radiochimique G (Fe(III)) dans ces conditions vaut 15,6 (3). La production d'ions ferriques après irradiation est mesurée par spectrophotométrie : les densités optiques (DO) des solutions irradiées et non irradiées sont comparées à λ = 304 nm, maximum d'absorption de l'ion Fe^{3+}. A cette longueur d'onde, le coefficient d'absorption molaire de l'ion Fe^{3+} est de 2204 cm^{-1} $mole^{-1}$ à 25°C.

1.d. Les spectres RPE

Les spectres des échantillons solides, irradiés à température ambiante, ont été enregistrés à cette température et à des températures plus élevées sur un appareil VARIAN E-12-bande X.

Les conditions d'enregistrement sont les suivantes :

. Champ	: 3334 Gauss
. Puissance	: 1 m Watt
. Modulation de champ	: 5 Gauss
. Fréquence	: 9,275 GHz
. Constante de temps	: 0,004 sec.

En vue d'étudier la cinétique de disparition des radicaux, la hauteur d'une raie déterminée du signal enregistré est mesurée en fonction du temps et de l'élévation de température. Cette hauteur est proportionnelle à la quantité d'espèces paramagnétiques présentes dans la cavité de résonance pour autant que la forme du spectre ne subisse pas de variation.

Aux températures supérieures à la température initiale, il est nécessaire de corriger la hauteur de la raie selon la relation de Kosirev [1] :

$$H_{T_0} = H_{Ti} \frac{T_i}{T_0} \quad (6)$$

H_{T_0} = hauteur de la raie corrigée

H_{Ti} = hauteur de la raie mesurée à la température T_i

T_0 = température de la première mesure soit 293 K

T_i = température de la mesure en degré K

2. RESULTATS

2.a. Le spectre

Les spectres des solides irradiés à deux doses différentes, à 298 K, sont représentés sur la figure 1.

La forme des spectres est inchangée. Seule la concentration radicalaire (hauteur des raies) augmente lorsque la dose est plus élevée.

Ces spectres sont caractéristiques des solides irradiés : ils attestent que la substance a été soumise à l'action de rayonnements ionisants puisqu'elle renferme des radicaux libres. En fait, un mélange de radicaux est produit et l'interprétation qualitative du spectre est délicate du fait de la trop grande largeur des raies.

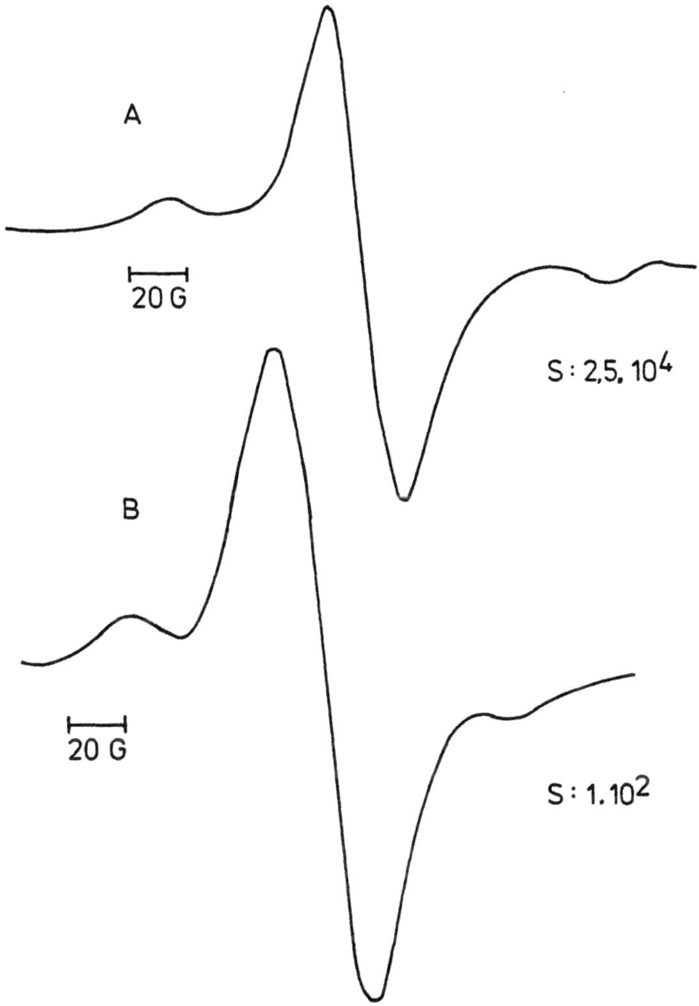

Figure 1 : Spectres RPE du diazépam irradié

A : irradié à 10,0 kGy, enregistré à 293 K
B : irradié à 381 kGy, enregistré à 293 K

2.b. Evolution de l'allure du spectre avec la température

Une augmentation de température provoque une recombinaison radicalaire. La figure 2 illustre ce phénomène.

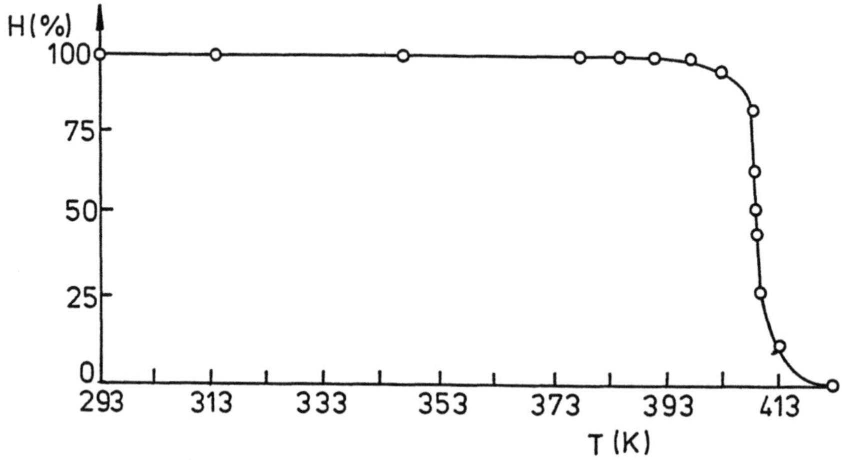

Figure 2 : décroissance radicalaire en fonction de la température pour DIAZ irradié à 18,8 kGy

Le pourcentage de radicaux résiduels dans le milieu diminue lorsque la température augmente pour s'annuler à une température proche de la fusion. Cette observation permet d'écarter toute transition solide-solide dans le domaine de température étudié ; une telle transition s'accompagnerait d'une perte de radicaux variable selon l'entropie de la transition ([1]).

2.c. Décroissance isotherme des radicaux piégés

A la température à laquelle une diminution significative des radicaux est observée, une cinétique de décroissance isotherme de décroissance radicalaire a pu être observée pour des échantillons irradiés à 2 doses différentes. Les courbes obtenues sont présentées sur la figure 3.

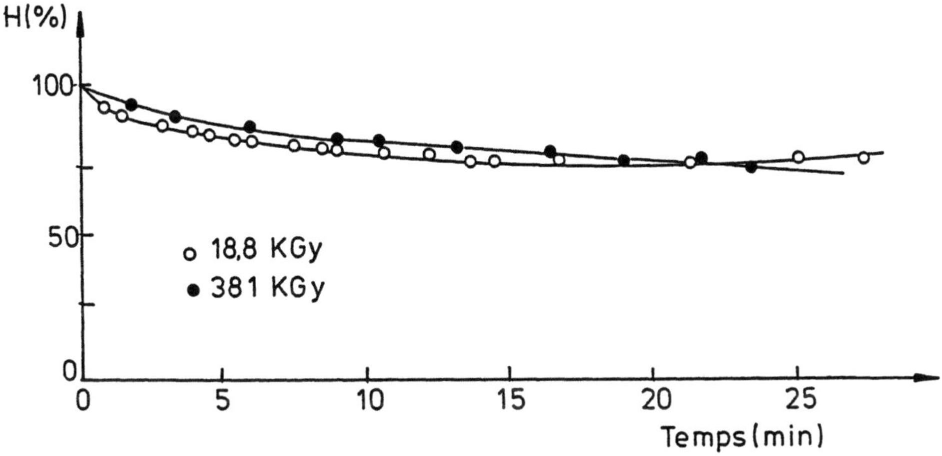

Figure 3 : Courbe de décroissance normalisée pour deux concentration initiales différentes en radicaux. Température d'observation : 403 K

Si on traite les données cinétiques par les équations correspondant aux ordres 1 et 2, le diagramme de kt en fonction de t aura l'allure de la figure 4.

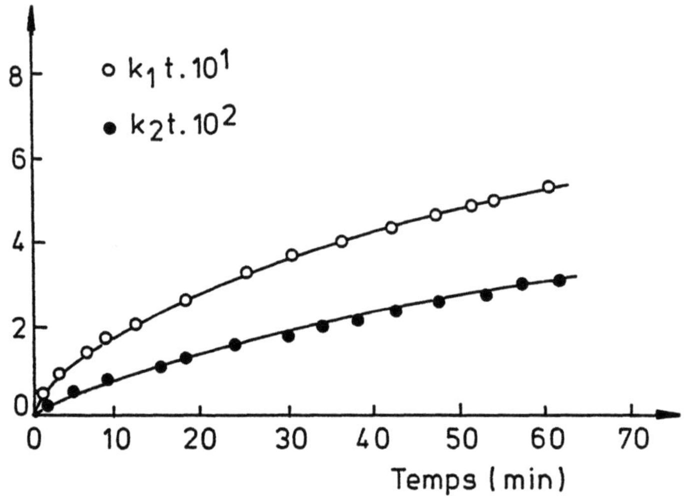

<u>Figure 4 : Traitement cinétique de la décroissance radicalaire isotherme des radicaux du Diaz. Dose : 381 kGy</u>

Aucun des 2 ordres ne donne une linéarité satisfaisante.

2.d. <u>Identification du ou des radicaux</u>

Comme précisé plus haut, les spectres enregistrés ne permettent pas de décrire la structure des radicaux libres présents dans ce solide.

Une méthode qui s'est avérée possible pour palier à ce manque est le "spin trapping" qui consiste à faire réagir, <u>en solution</u>, le radical piégé avec un composé appelé le "spin-trap". Ce dernier est une molécule capto-dative capable donc de stabiliser le radical produit en solution.

Les essais ont été réalisés sur le DIAZ irradié à 381 kGy.

Le spin-trap utilisé est représenté ci-dessous : il s'agit du α-t-butylmercaptoacrylonitrile (TBMA).

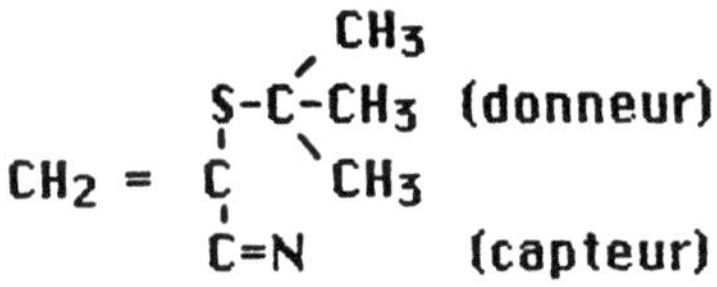

(TBMA)

Le phénomène de spin-trapping est résumé dans les réactions chimiques suivantes[4]:

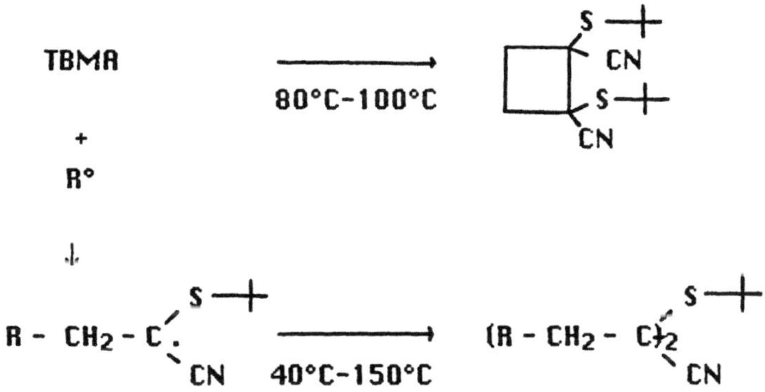

(1)

Le radical obtenu (1) dimérise à température ambiante. Une augmentation de la température, de 40°C à 150°C selon la nature de R°, détruit le dimère de telle façon que le radical (1) est observé.

Le solvant est le dichlorobenzène. Son caractère aromatique évite que le radical ne réagisse avec le solvant.

Le spectre obtenu est reproduit sur la figure 5.

Conditions : . Champ : 3231 G
. Puissance : 1,5 mWatt
. Modulation de champ : 1 Gauss
. Fréquence : 9,17 GHz
. Constante de temps : 2 secondes

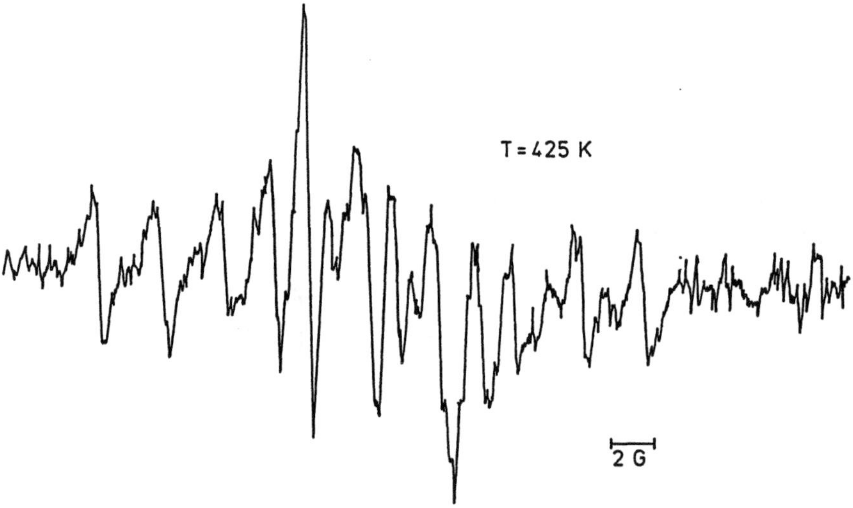

Figure 5 : Spectre RPE du radical obtenu par réaction du TBMA sur DIAZ irradié à 381 kGy. Température 152°C

L'allure du spectre serait expliqueé par la présence de deux radicaux différents dont un serait plus abondant. Sa structure est reproduite ci-dessous :

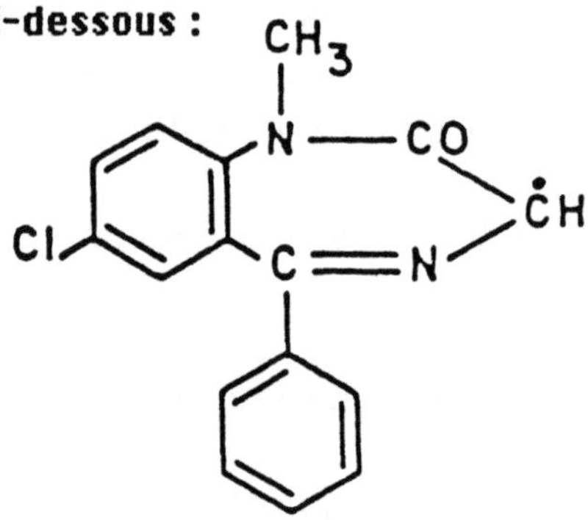

3. Discussion de la cinétique radicalaire

En phase solide, le transfert de l'énergie ou de la charge peut être observé par des distances correspondant à de nombreux diamètres moléculaires ; en outre, la phase solide affecte aussi les réactions chimiques en piégeant les espèces radicalaires. Par rapport à une irradiation en phase liquide, on constate une diminution des rendements radiochimiques G due, pour l'essentiel, à une dissipation de l'énergie cédée. La décroissance radicalaire en cours de radiolyse est attribuée à la recombinaison de radicaux formés en des sites voisins dans une même grappe : le partenaire n'est pas choisi au hazard. Mais la plupart des radicaux sont piégés et diffusent dans le milieu après la radiolyse lorsque la température est augmentée. Il s'en suit une diminution de la concentration en radicaux dans le milieu à laquelle est associée une cinétique.

Cette cinétique de décroissance radicalaire est caractéristique de la radiolyse : la distribution des radicaux en zones hétérogènes devrait s'accompagner d'une cinétique d'un ordre 1 par rapport à la dose cédée :

$$V = k\,[\text{grappe}] = k\,[\text{dose}] \qquad (7)$$

Dans ce cas, les courbes de décroissance radicalaire isotherme, pour des irradiations à deux doses différentes peuvent se superposer après normalisation : les radicaux d'une même grappe réagissent entre eux (cinétique inhomogène). Dans le cas contraire, les radicaux de grappes différentes réagissent entre eux :

$$V = k\,[\text{grappe}]^2 = k\,[\text{Dose}]^2 \qquad (8)$$

Les courbes de décroissance radicalaire ne seront pas superposables puisque la diminution est proportionnelle au carré de la concentration en radicaux.

En pratique, les radicaux réagissent avec des partenaires qui ne sont pas nécesairement issus de la même grappe [5] mais la cinétique n'est toutefois pas entièrement homogène. Par un traitement cinétique classique (ln C et 1/C en fonction du temps), une succession d'ordres 2 pourra être observée. Toutefois, si les impuretés présentes dans le milieu étaient à l'origine des radicaux piégés, la cinétique de décroissance

décroissance radicalaire serait d'ordre 2. En effet, lors de la radiolyse, le composé principal S reçoit l'énergie cédée et les impuretés I (inclusions) peuvent être l'objet d'un transfert d'énergie ou de charge. On parle alors d'un transfert de l'activation chimique :

$$S^* + I \longrightarrow S + I^*$$

La distribution des radicaux dus aux impuretés est quasi homogène et la décroissance radicalaire n'est plus caractéristique de la radiolyse.

Application au diazépam

Les courbes de décroissance radicalaire en fonction du temps (figure 3) sont proches l'une de l'autre. Leur superposition indique une proportionnalité entre la vitesse de disparition des radicaux et la dose cédée. Les radicaux d'une même grappe réagissent entre eux.

Aucun ordre ne correspond à cette décroissance. Elle peut être traitée par une succession d'ordre 2.

Ces deux constatations amènent à penser que seule la molécule de diazépam subit les effets des rayons γ du ^{60}Co.

Conclusions

La RPE se révèle être une technique de choix pour démontrer qu'un solide moléculaire a été irradié, ce qui n'est pas le cas d'autres techniques telles le point de fusion, le pH, la chromatographie...

De plus, elle permet l'étude de la répartition des radicaux libres issus de l'irradiation.

(1) B.Tilquin
"Composante radicalaire des transformations chimiques radioinitiées dans les alcanes à 77 K."
Ed.Cabay, Louvain-la-Neuve, 1985.

(2) J.E.Willard
dans "Radiation and Chemistry : Principles and Applications."
Chap.13, Ed.Farhataziz et Michaël A.J.Rodgers.
New-York, VCH, 1987.

(3) C.Ferradini et J.Pucheault
"Biologie de l'action des rayonnements ionisants."
Ed.Masson, 1983.

(4) S.Mignani ; M.Beaujean ; Z.Janousek ; R.Merenyi et H.G.Viehe
Tetrahedron, vol.37, Suppl.n°1, pp.111-115 (1981).

(5) J.E.Willard
"Fundamental process in radiation chemistry".
Chap.9, Ed.Ausloos, 1968.

MIX
Papier aus verantwortungsvollen Quellen
Paper from responsible sources
FSC® C105338

If you have any concerns about our products,
you can contact us on
ProductSafety@springernature.com

In case Publisher is established outside the EU,
the EU authorized representative is:
Springer Nature Customer Service Center GmbH
Europaplatz 3, 69115 Heidelberg, Germany

Printed by Libri Plureos GmbH
in Hamburg, Germany